STUDENT'S SOLUTION MANUAL TO ACCOMPANY

INTRODUCTORY STATISTICS

Second Edition

Jay Devore
California Polytechnic State University
San Luis Obispo

Roxy Peck
California Polytechnic State University
San Luis Obispo

Prepared by
John E. Groves
California Polytechnic State University
San Luis Obispo

West Publishing Company

Minneapolis/St. Paul New York Los Angeles San Francisco

WEST'S COMMITMENT TO THE ENVIRONMENT

In 1906, West Publishing Company began recycling materials left over from the production of books. This began a tradition of efficient and responsible use of resources. Today, up to 95% of our legal books and 70% of our college texts and school texts are printed on recycled, acid-free stock. West also recycles nearly 22 million pounds of scrap paper annually—the equivalent of 181,717 trees. Since the 1960s, West has devised ways to capture and recycle waste inks, solvents, oils, and vapors created in the printing process. We also recycle plastics of all kinds, wood, glass, corrugated cardboard, and batteries, and have eliminated the use of Styrofoam book packaging. We at West are proud of the longevity and the scope of our commitment to the environment.

Production, Prepress, Printing and Binding by West Publishing Company.

610 Opperman Drive
P.O. Box 64526
St. Paul, MN 55164–0526

Printed in the United States of America
00 99 98 97 96 95 94   8 7 6 5 4 3 2

ISBN 0–314–03499–4

CONTENTS

Chapter 1 Introduction 1

Chapter 2 Tabular and Pictorial Methods for Describing Data 3

Chapter 3 Numerical Summary Measures 23

Chapter 4 Probability 35

Chapter 5 Random Variables and Discrete Probability Distributions 49

Chapter 6 Continuous Probability Distributions 61

Chapter 7 Sampling Distributions 71

Chapter 8 Estimation Using a Single Sample 81

Chapter 9 Hypothesis Testing Using a Single Sample 89

Chapter 10 Comparing Two Populations or Treatments 107

Chapter 11 Regression and Correlation: Descriptive Methods 129

Chapter 12 Regression and Correlation: Inferential Methods 147

Chapter 13 The Analysis of Variance 173

Chapter 14 Categorical Data and Goodness-of-Fit Tests 185

PREFACE

This solutions manual accompanies the text *INTRODUCTORY STATISTICS,* Second Edition by Jay L. Devore and Roxy L. Peck. It contains solutions to all the odd numbered problems in the text.

The person using this manual should be aware that they might obtain slightly different numerical answers than those given in the manual. This difference can occur when a rounding procedure different from that in the solution is used. This difference may manifest itself particularly in chapters 11, 12, and 13. It is suggested that the user maintain as many decimal places as possible in doing intermediate calculations.

Even though I have checked the solutions carefully, it is possible that errors are still present. I would appreciate having any remaining errors brought to my attention.

Finally, I would like to acknowledge the patience shown and the support given by my family during the period of this project. In particular, I would like to express my sincere admiration and thanks to my daughters, Alicia and Heather, for the excellent job they did typing the manuscript.

J. E. G.

Chapter 1
Introduction

Section 1.2

1.1 Descriptive statistics is made up of those methods whose purpose is to organize and summarize a data set. Inferential statistics refers to those procedures or techniques whose purpose is to generalize or make an inference about the population based on the information in the sample.

1.3 The population of interest is the entire student body (the 15,000 students). The sample consists of the 200 students interviewed.

1.5 The population consists of all single family homes in Pasadena. The sample consists of the 100 homes selected for inspection.

1.7 The population consists of all 5000 bricks in the lot. The sample consists of the 100 bricks selected for inspection.

Section 1.3

1.9 Since each case has a case number, write each case number on a slip of paper; one case number per slip of paper. Place the 870 slips of paper into a container and thoroughly mix the slips. Then, select 50 slips (one at a time, without replacement) from the container. The 50 cases whose case numbers are on the 50 selected slips constitute the random sample of 50 cases.

1.11 Stratified sampling would be worthwhile if, within the resulting strata, the elements are more homogeneous than the population as a whole.

a) As one's class standing increases, the courses become more technical and therefore the books become more costly. Also, there might be fewer used books available. Therefore, the amount spent by freshmen might be more homogeneous than the population as a whole. The same statement would hold for sophomores, juniors, seniors, and graduate students. Therefore, stratifying would be worthwhile.

b) The cost of books is definitely dependent on the field of study. The cost for engineering students would be more homogeneous than the general college population. A similar statement could be made for other majors. Therefore, stratifying would be worthwhile.

c) There is no reason to believe that the amount spent on books is connected to the first letter in the last name of a student. Therefore, it is doubtful that stratifying would be worthwhile.

1.13 The response variable for the tile is whether it cracked or not in the firing process. Since two different firings will not have exactly the same temperature, tiles made from each type of clay should be fired together. Fifty tiles of each type could be used per firing. Since temperature varies within the oven, the oven should be divided into sections (blocks) where the temperature is the same within a section, but perhaps differs between sections. Then, files made from each clay type should be placed within the sections. The positions of the tiles within a section should be determined in a random fashion.

Chapter 2
Tabular and Pictorial Methods for Describing Data

Section 2.1

2.1
- a) numerical (discrete)
- b) categorical
- c) numerical (continuous)
- d) numerical (continuous)
- e) categorical

2.3
- a) discrete
- b) continuous
- c) discrete
- d) discrete
- e) continuous
- f) continuous
- g) continuous
- h) discrete

Section 2.2

2.5
- a) 2.2 liters/min.
- b) In the row with stem of 8. The leaf of 9 would be placed to the right of the other leaves.
- c) A large number of flow rates are between 6.0 and 8.0. Perhaps 6.9 or 7.0 could be selected as a typical flow rate.
- d) There appears to be quite a bit of variability in the flow rates. While there is a large number of flow rates in the 6.0 to 9.0 range, the flow rates appear to vary quite a bit in relation one to another.
- e) The distribution is not symmetric. Taking 7.0 as a typical value, the smaller flow rates are spread from 2.2 to 7.0, while the larger flow rates are spread from 7.0 to 18.9 (a larger spread).
- f) The value 18.9 appears to be somewhat removed from the rest of the data and hence is outlier.

2.7

Stem	Leaf
1H	8
2L	0,2,4
2H	9,8,7,5,6,9,7,8,5,8,5,5,8,7,7,8,7,9,6,6,7,5,8,5
3L	1,4
3H	6
4L	
4H	7,9
5L	4

stem: ones
leaf : tenths

2.9

Stem	Leaf
0	6,7,5,8,9,0,0,5,0
1	8,6,1,5,2,1,6,7,0,6,3,1,2,2,8,4,3,1,8,0,8,7,8,7,7
2	5,6,9,0,7,8,6,9,0,6,1,9,7,1
3	7,5,5,0,0,0,1,5,0,5
4	5,6,7,8,2,3,0
5	
6	7,1
7	0

stem: tens digit
leaf: ones digit

The data values are concentrated between 0 to 40, with a few larger values. The values concentrated between 0 and 40 appear to be distributed almost symmetrically.

2.11

Creamy		**Crunchy**
2	2	
0,0,9,6	3	4,4,6
0,1,0,5,4	4	2,7,0,2,7
6,0,6,0,3,6	5	3,2,0,6
8,5,2	6	2,2,2
	7	5,5
	8	0

stem: tens
leaf: ones

The scores for Crunchy tend to be higher than those for Creamy. The scores for the two groups have similar variability and spread, but the typical score for Crunchy is larger than the typical score for Creamy.

2.13

Athletes	Stem	Nonathletes
	0L	3,3,4
8,8,9	0H	6,7,8,9
0,0,0,2,2,3,3,4,4	1L	2,3,3,4,4,4,4,4
5,5,5,5,5,6,6,6,7,7,7,7,8,8,9	1H	5,5,5,6,7,7,7,8,8,8,8
0,1,1,1,1,3,4	2L	0,2,3,4,4
6,7,7,8,9,9	2H	5,5,5,8,9,9
	3L	0,4,4,4
5,5,6,7,7,9	3H	8
0	4L	4
	4H	5,8,8
	5L	2

stem: tens
leaf: ones

Based on the stem and leaf displays, there does not seem to be any evidence that the proportion of students disqualified is smaller for nonathletes than for athletes. There appears to be more variation for nonathletes, but other than that, both displays have generally the same shape and location.

Section 2.3

2.15

Categories	Frequency	Relative Frequency
R	16	.32
M	15	.30
C	12	.24
S	7	.14
	n = 50	1.00

2.17 a)

Category	Frequency	Relative Frequency
1	1	.0059
2	2	.0118
3	13	.0765
4	19	.1118
5	35	.2059
6	38	.2235
7	33	.1941
8	18	.1059
9	8	.0471
10	2	.0118
11	1	.0059
	n = 170	1.0002 $\approx$ 1.0

b) The proportion of litters that were greater than 6 in size was .1941 + .1059 + .0471 + .0118 + .0059 = 0.3648. The proportion of litters that were between 3 and 8 inclusively was .0765 + .1118 + .2059 + .2235 + .1941 + .1059 = 0.9177.

c) Yes. It is considerably easier to answer questions like that posed in (b) using the relative frequency distribution, because the data has been categorized. If it weren't, then for each question it would be necessary to go through the raw data searching for those observations pertinent to the question being answered.

2.19 a)

Number of Fish Caught	**Number of Anglers**	**Relative Frequency**	**Cumulative Relative Frequency**
0	515	.5653	.5653
1	65	.0714	.6367
2	60	.0659	.7026
3	66	.0724	.7750
4	53	.0582	.8332
5	55	.0604	.8936
6	27	.0296	.9232
7	25	.0274	.9506
8	25	.0274	.9780
9	20	.0220	1.0000
	n = 911	1.0000	

b) The proportion of anglers in the sample who caught no fish is .5653. The proportion of anglers in the sample who caught one fish is .0714. The proportion of anglers in the sample who caught at most one fish is .5653 + .0714 = .6367.

c) The proportion of anglers in the sample who caught at least five fish is .0604 + .0296 + .0274 + .0274 + .0220 = .1668. The proportion of anglers in the sample who caught more than five fish is .0296 + .0274 + .0274 + .0220 = .1064.

d) See part (a) for the cumulative relative frequencies. The proportion of anglers in the sample who caught at least 5 fish is 1 − .8332 = .1668. The proportion of anglers in the sample who caught more than 5 fish is 1 − .8936 = .1064.

e)

Number of Fish Caught	Number of Anglers (Frequency)
0	515
1	65
2	60
3	66
4	53
5	55
6	27
7	25
8	25
9	20
10 or more	4
	n = 915

2.21 a) The suggested class intervals would be inappropriate because there are observations whose values are 30.0 and 40.0. Therefore, there is uncertainty in which intervals these observations should be placed. Does 30.0 get placed into 20-30 or 30-40? Does 40.0 get placed into 30-40 or 40-50? Secondly, there are no observations below 20, so 0-10 and 10-20 are empty classes.

b)

Concentration	Frequency
20 -< 30	1
30 -< 40	8
40 -< 50	8
50 -< 60	6
60 -< 70	16
70 -< 80	7
80 -< 90	2
90 -< 100	2
	n = 50

c) The proportion of the concentration observations that were less than 50 is

$$\frac{(8+8+1)}{50} = \frac{17}{50} = .34$$

The proportion of the concentration observations that were at least 60 is

$$\frac{(16+7+2+2)}{50} = \frac{27}{50} = .54$$

d) No, because of the 16 observations in the class 60 -< 70, there is no information in the frequency distribution to indicate how many are less than 65. To estimate the proportion one could assume that 8 of the 16 were between 60 -< 65. Hence, half of the observations in the 60 -< 70 class are in the lower half, 60 -< 65. Thus, an estimate of the proportion of observations in the sample which are less than 65 would be

$$\frac{(1+8+8+6+8)}{50} = \frac{31}{50} = .62$$

The actual proportion (from the original data) is $\frac{31}{50}$ = .62.

2.23 a) No. If a small number of classes are used, then the resulting classes would be very wide with most observations in one or two classes. If a large number of classes are used, then many of the classes would have zero frequencies.

b)

Class Intervals	Frequency	Relative Frequency
5 -< 10	11	.22
10 -< 15	10	.20
15 -< 20	8	.16
20 -< 25	8	.16
25 -< 30	2	.04
30 -< 40	5	.10
40 -< 50	2	.04
50 -< 100	3	.06
100 -< 500	1	.02
	n = 50	1.00

2.25 a) and b)

Classes	Frequency	Relative Frequency	Cum. Rel. Frequency
0 -< 6	2	.0225	0.0225
6 -< 12	10	.1124	0.1349
12 -< 18	21	.2360	0.3709
18 -< 24	28	.3146	0.6855
24 -< 30	22	.2472	0.9327
30 -< 36	6	.0674	1.0001
	n = 89	1.0001	

c) (Rel. Freq. for 12 -< 18) = (Cum. Rel. Freq. for < 18) – (Cum. Rel. Freq. for < 12) = .3709 – .1349 = .2360.

d) The proportion that had pacemakers that did not malfunction within the first year equals 1 minus the proportion that had pacemakers that malfunctioned within the first year (12 months), which is 1 – .1349 = .8651, which converts to 86.51%.

e) The proportion that required replacement between one and two years after implantation is equal to the proportion that had to be replaced within the first 2 years (24 months) minus the proportion that had to be replaced within the first year (12 months). This is 0.6855 – 0.1349 = 0.5506, which converts to 55.06%.

f) The proportion that lasted less than 18 months is .3709, which converts to 37.09%, and the proportion that lasted less than 24 months is .6855, which converts to 68.55%. Thus, the time at which about 50% of the pacemakers failed is somewhere between 18 and 24 months. A more precise estimate can be found as follows.

$$\frac{(.50 - .3709)}{.3146} = \frac{x}{6} \Rightarrow x = \frac{6(.50 - .3709)}{.3146} = 2.46$$

So the time at which about 50% of the pacemakers had failed is 18 + 2.46 or 20.46 months.

g)

$$\frac{(.9327 - .9)}{.2472} = \frac{x}{6} \Rightarrow x = \frac{6(.9327 - .9)}{.2472} = .79$$

So an estimate of the time at which only 10% of the pacemakers initially implanted were still functioning is 30 – .79 = 29.21 months.

2.27 It would not make sense to calculate cumulative relative frequencies for the data since it is categorical data.

Section 2.4

2.29 a)

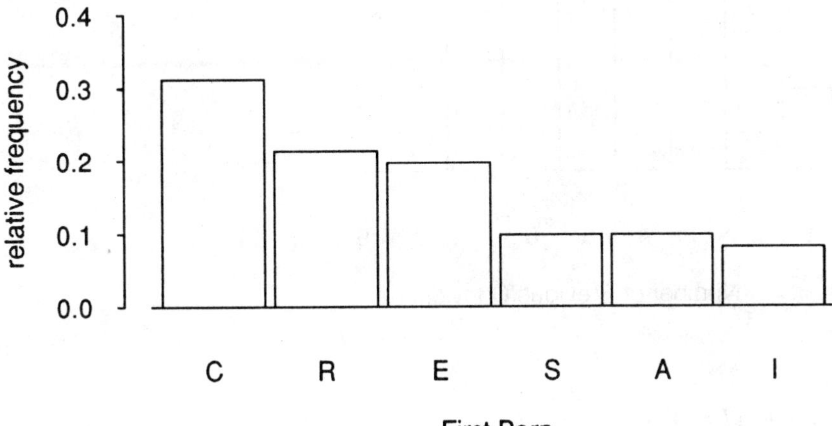

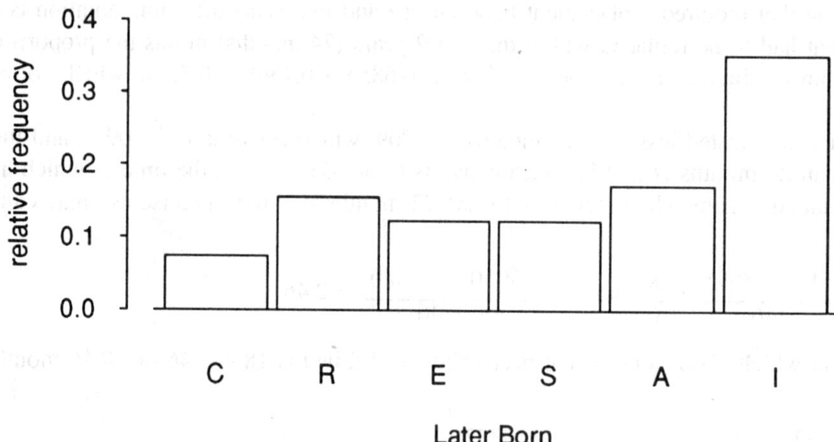

b) The occupational preferences of first-born differ from those of later born. First-born are most likely to be conventional, while later born are most likely to be investigative.

2.31

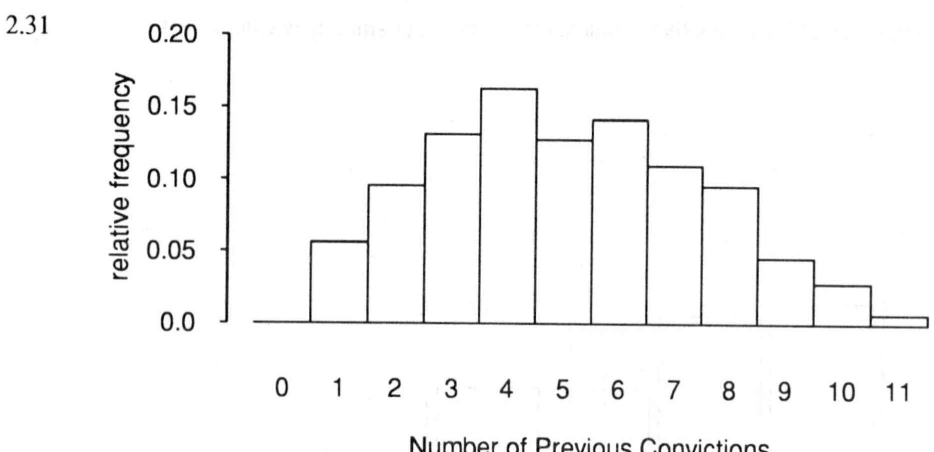

2.33 a) .08

.08 + .115 + .11 + .17 = .475

.025 + .01 + .01 + .01 = .055

b) The waiting times are concentrated in the 2 minutes to 7 minutes range, with some waiting times as large as 20 minutes. The most prevalent waiting times are around 5 minutes. The histogram is positively skewed.

2.35 a) The height of the rectangle appears to be greater than .16, about .163. The relative frequency for the 6 -< 8 class would be .163(2) = .326.

b) About .047(2) + .097(2) = .094 + .194 = .288
About .034(5) + .008(5) = .170 + .040 = .210

Comment to the student: Your answers may vary somewhat from the answers above, because the heights of the rectangles that you read from the graph may differ from my readings of the heights.

2.37 a) and b)

Class	Class Width	Rel. Freq.	Height = Rel. Freq./Class Width
1	40	17/191 = .0890	.0022
2	20	16/191 = .0838	.0042
3	20	25/191 = .1309	.0065
4	20	34/191 = .1780	.0089
5	20	46/191 = .2408	.0120
6	20	33/191 = .1728	.0086
7	40	18/191 = .0942	.0024
8	40	2/191 = .0105	.0003

c)

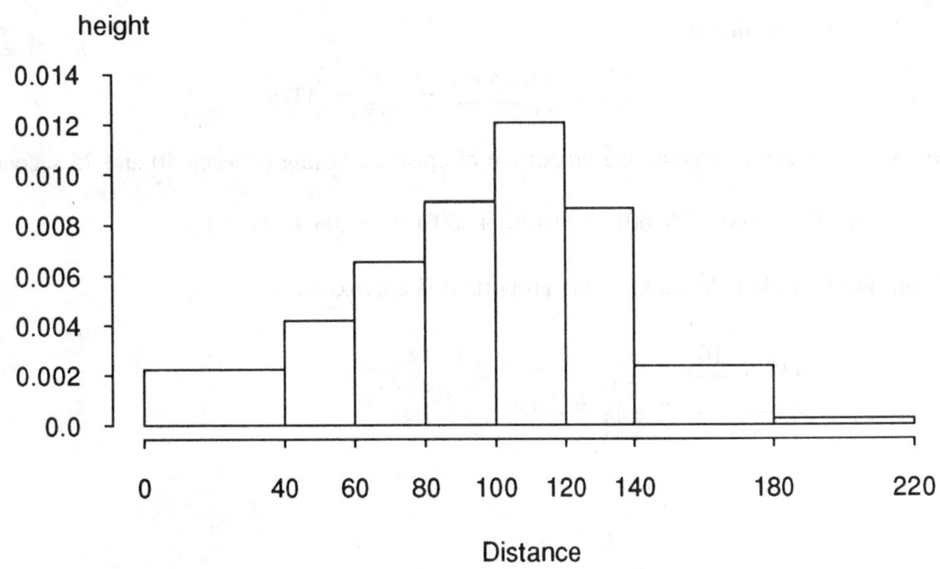

d) The area of the rectangle above the fifth class interval is 20(.012) = .2408 (the rel. freq.). The total area of all rectangles in the histogram is the sum of the relative frequencies, which is 1.0.

2.39 a)

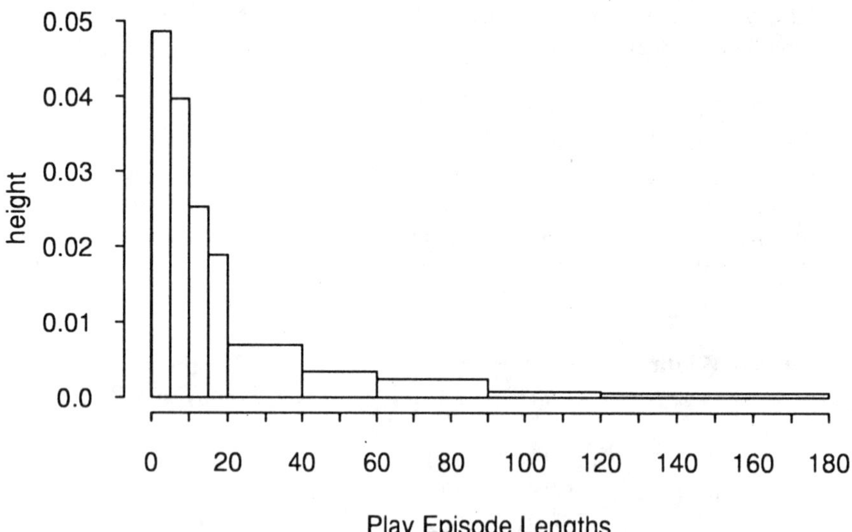

b) From the density histogram, the proportion of episodes lasting at least 20 seconds is approximately

$.008(40-20) + .003(60-40) + .002(90-60) + .001(120-90) + .0005(180-120)$

$= .008(20) + .003(20) + .002(30) + .001(30) + .0005(60)$

$= .16 + .06 + .06 + .03 + .03$
$= .34$

The actual proportion is

$$\frac{(31+15+16+5+8)}{222} = \frac{75}{222} = .3378$$

c) From the density histogram, the proportion of episodes lasting between 40 and 75 seconds is approximately

$.003(60-40) + (.002)(75-60) = .003(20) + .002(15) = .06 + .03 = .09$

From the frequency distribution, the proportion is approximately

$$\frac{(15 + \frac{16}{2})}{222} = \frac{23}{222} = .1036$$

2.41

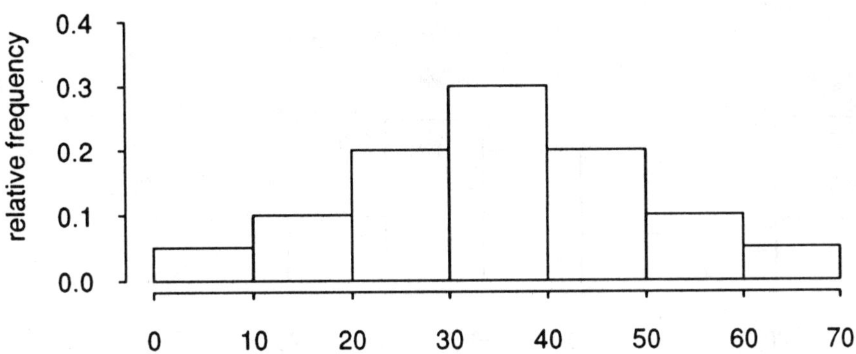

This histogram is symmetric.

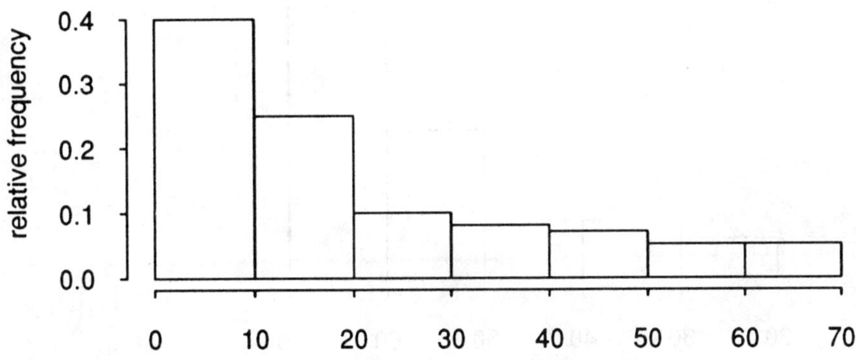

This histogram is positively skewed.

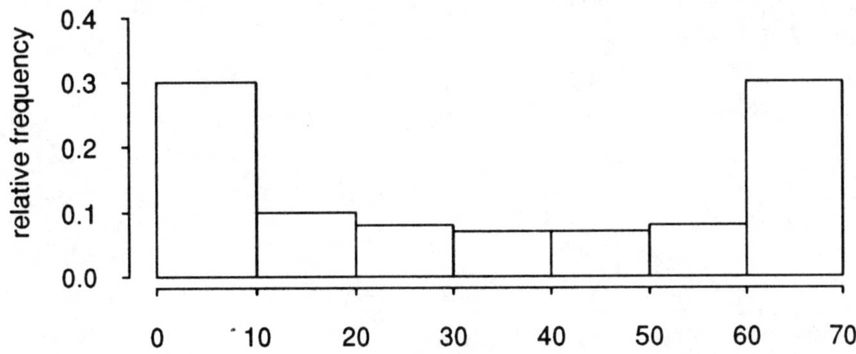

This is a bimodal histogram. While it is not perfectly symmetric it is close to being symmetric.

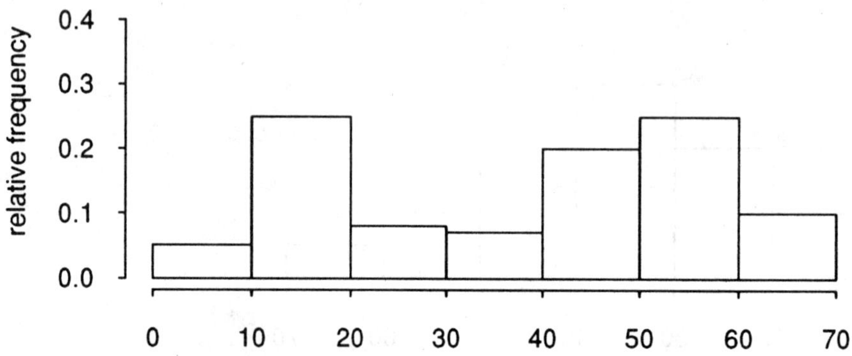

This is a bimodal histogram.

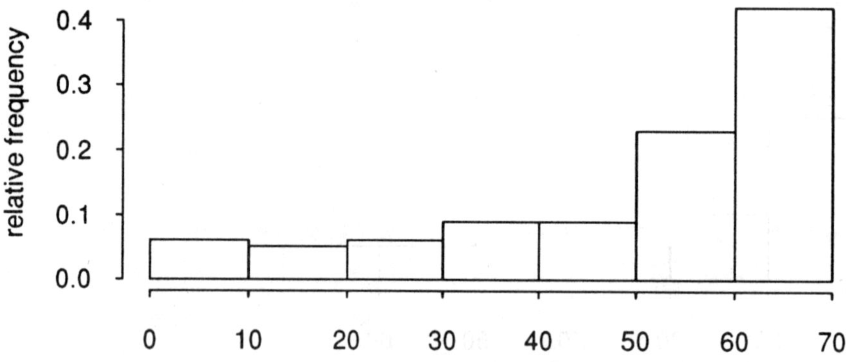

This is a negatively skewed histogram.

Supplementary Exercises

2.43

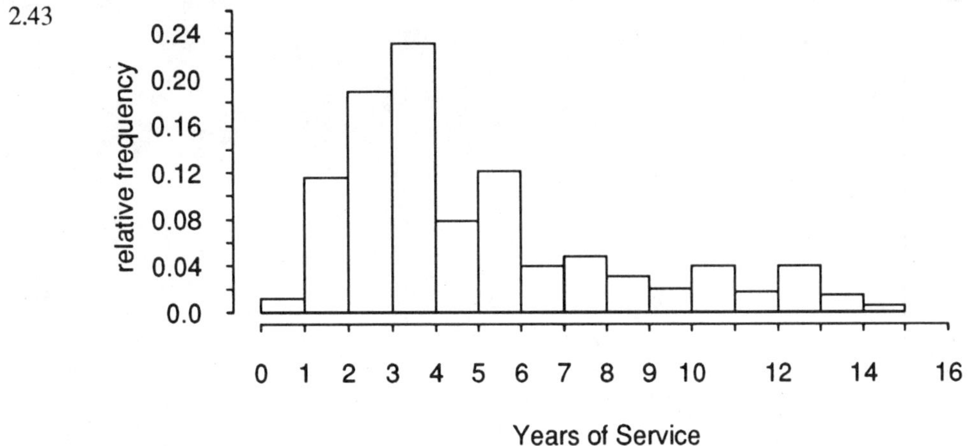

This histogram is positively skewed.

2.45 a)

Work Status of Mother	Relative Frequency
Unemployed	.6833
Employed Part Time	.2167
Employed Full Time	.1000
	1.0000

The proportion of mothers who work outside the home is .2167 + .1000 = .3167 or 31.67%.

b)

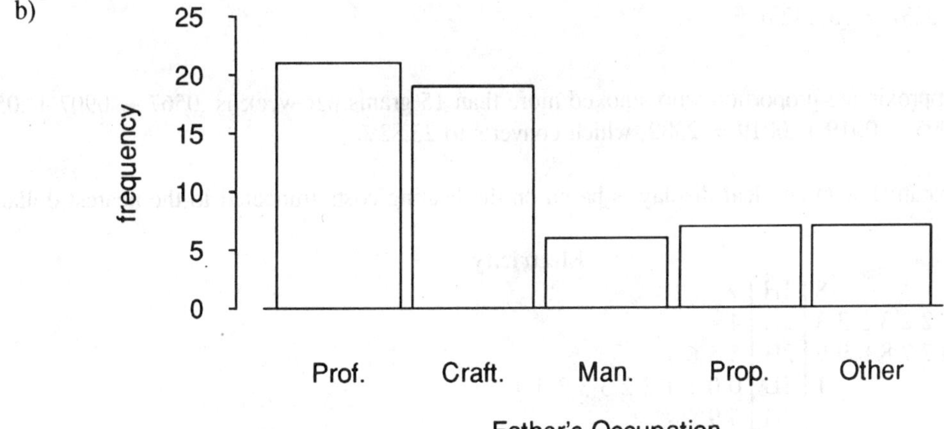

2.47 a)

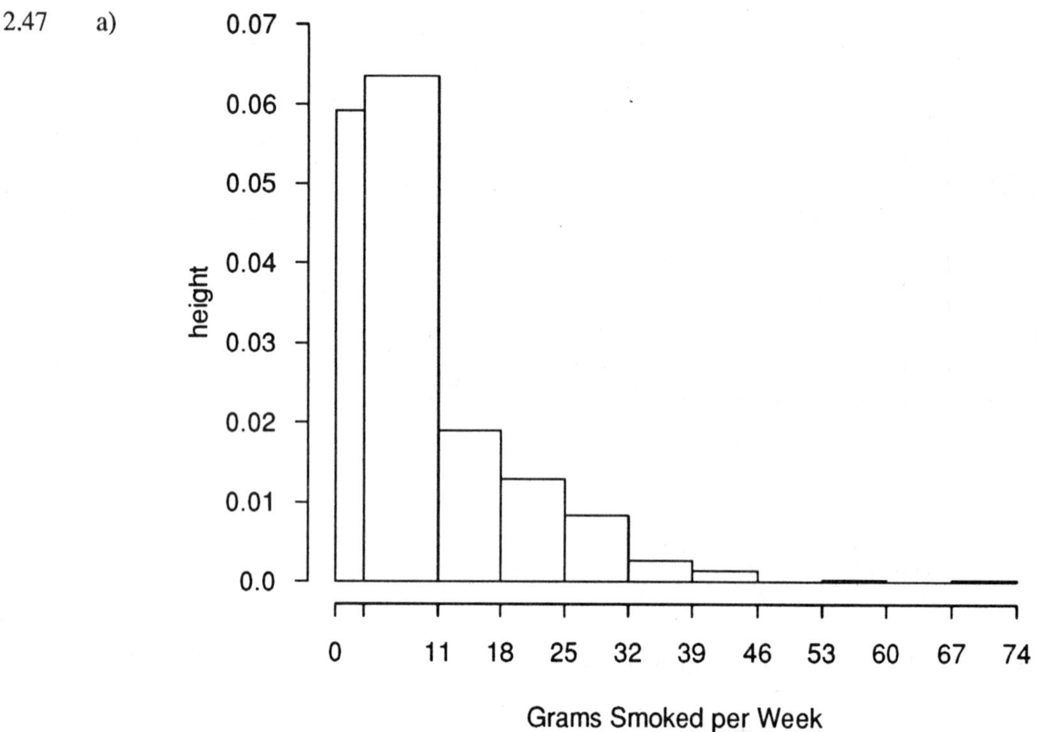

b) The proportion of respondents who smoked 25 or more grams per week is $\frac{48}{529}$, or using the relative frequencies .0586 + .0189 + .0095 + .0019 + .0019 = .0908, which converts to 9.08%.

c) The approximate proportion who smoked between 15 and 18 grams per week is

$$\frac{(18-15)}{(18-11)}(.1323) = \frac{3}{7}(.1323) = .0567$$

Thus, the approximate proportion who smoked more than 15 grams per week is .0567 + .0907 + .0586 + .0189 + .0095 + .0019 + .0019 = .2382, which converts to 23.82%.

2.49 The following comparative stem and leaf display is based on the heating costs truncated to the nearest dollar.

Gas		Electricity
8	1H	9
0 1 2 2 3 3 3 3	2L	4
5 5 5 5 6 6 6 6 6 7 7 8 8 9 9	2H	5 5 6 7
1	3L	0 0 1 1 2 2 3 3 3 4 4
	3H	5 9
	4L	0 1 3 4
	4H	8
	5L	1

stem: tens
leaf: ones

The data suggests that the cost of heating a two-bedroom apartment in Southern California is considerably less expensive using gas than using electricity.

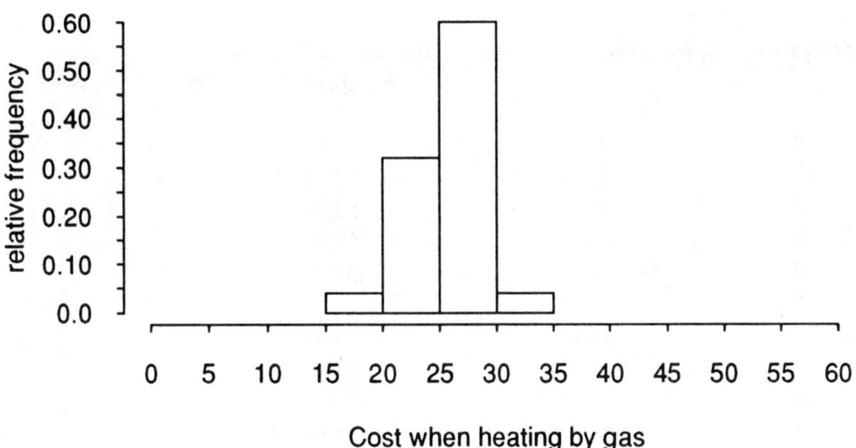
0.60
0.50
0.40
0.30
0.20
0.10
0.0
relative frequency
0 5 10 15 20 25 30 35 40 45 50 55 60
Cost when heating by gas

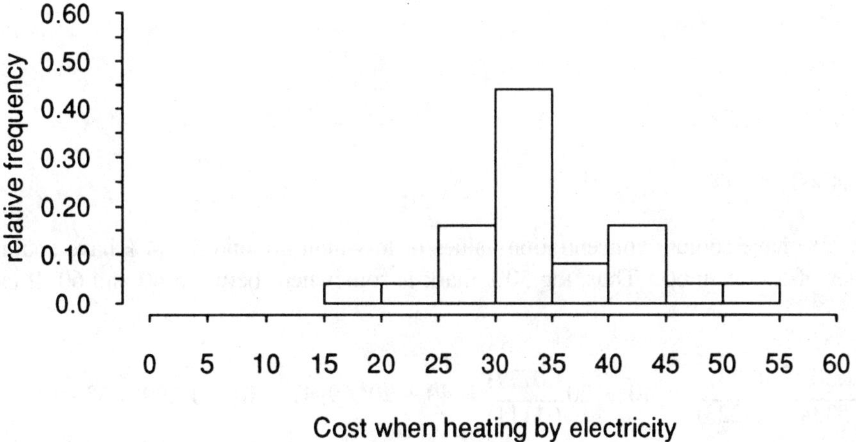
0.60
0.50
0.40
0.30
0.20
0.10
0.0
relative frequency
0 5 10 15 20 25 30 35 40 45 50 55 60
Cost when heating by electricity

2.51

Sodium Concentration(mM)	Frequency	Relative Frequency	Cumulative Relative Frequency
0 -< 20	7	.1556	.1556
20 -< 40	12	.2667	.4223
40 -< 60	5	.1111	.5334
60 -< 80	1	.0222	.5556
80 -< 100	0	.0000	.5556
100 -< 120	1	.0222	.5778
120 -< 140	1	.0222	.6000
140 -< 160	4	.0889	.6889
160 -< 180	8	.1778	.8667
180 -< 200	3	.0667	.9334
200 -< 220	2	.0444	.9778
220 -< 240	1	.0222	1.0000
	n = 45	1.0000	

a) .5556 or 55.56%

b) .9334 − .5556 = .3778 or 37.78%

c) 1.0000 − .6000 = .4000 or 40%

d) Approximately 42.23% have sodium concentration values of less than 40, and 53.34% have sodium concentration values of less than 60. Thus, the 50% mark is somewhere between 40 and 60. It can be approximated by

$$40 + (60 - 40)\frac{(.50 - .4223)}{(.5334 - .4223)} = 40 + 20\frac{(.0777)}{(.1111)} = 40 + 20(.6994) = 40 + 13.99 = 53.99$$

Thus, approximately half of the observed sodium concentrations are smaller than 53.99 (or 54).

2.53 A stem and leaf display for the data results in

```
3L | 1 2 3 3 3 4 4 4
3H | 5 7 7 9 9
4L | 0 0 0 0 0 1 2 3 3 4 4
4H | 5 5 5 6 6 6 6 6 6 6 7 7 7 7 7 7 8 8 8 8 8 8 9
5L | 0 0
```

where a stem of 3 and a leaf of 1 denote a number between 310 and 319. A frequency distribution for the data is:

Classes	Frequency	Relative Frequency
300 -< 320	1	.0204
320 -< 340	4	.0816
340 -< 360	4	.0816
360 -< 380	2	.0408
380 -< 400	2	.0408
400 -< 420	6	.1224
420 -< 440	3	.0612
440 -< 460	5	.1020
460 -< 480	13	.2653
480 -< 500	7	.1429
500 -< 520	2	.0408
	n = 49	.9998

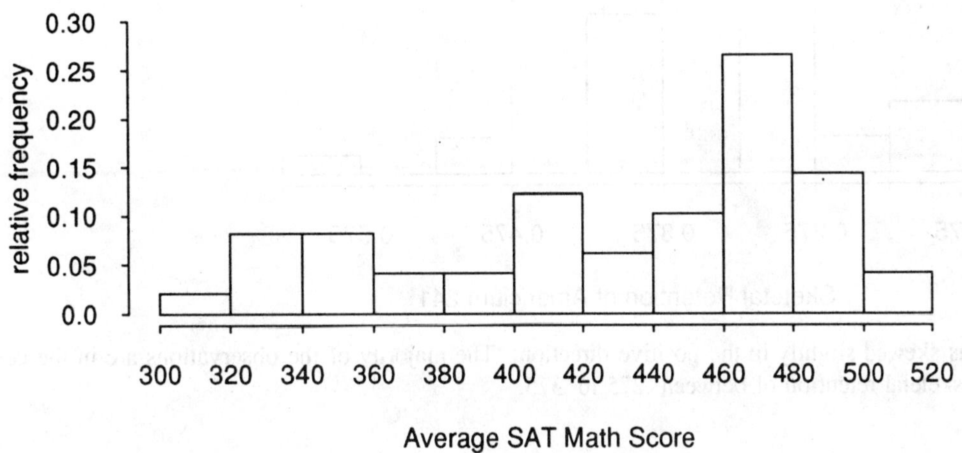

The scores varied from 317 to 503, with most schools scoring in the mid-400's.

2.55 a)

Class	Frequency	Relative Frequency
.175 -< .225	4	.0727
.225 -< .275	2	.0364
.275 -< .325	16	.2909
.325 -< .375	15	.2727
.375 -< .425	9	.1636
.425 -< .475	6	.1091
.475 -< .525	2	.0364
.525 -< .575	0	.0000
.575 -< .625	1	.0182
	n = 55	1.0000

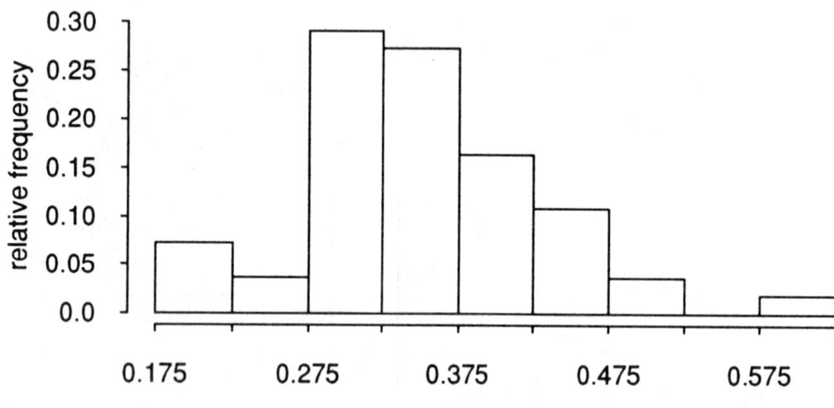

The histogram is skewed slightly in the positive direction. The majority of the observations are in the center. Most beagles have a skeletal retention of between .275 to .375.

2.57

Number of Publications	Frequency	Relative Frequency	Height
0	104	.2946	.2946
1 - 2	60	.1700	.0850
3 - 4	44	.1246	.0623
5 - 6	30	.0850	.0425
7 - 10	30	.0850	.0213
11 - 15	26	.0737	.0147
16 - 25	31	.0878	.0088
26 - 50	28	.0793	.0032
	n = 353	1.0000	

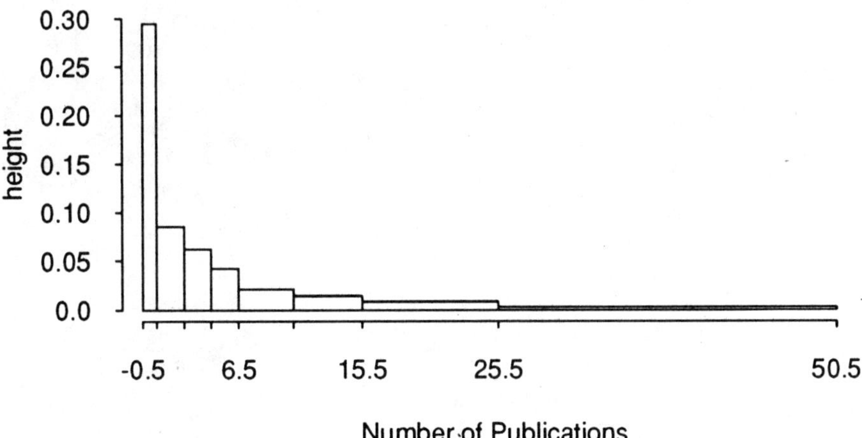
0.30
0.25
0.20
0.15
0.10
0.05
0.0
height
-0.5
6.5
15.5
25.5
50.5
Number of Publications

Chapter 3
Numerical Summary Measures

Section 3.1

3.1 a) $\bar{x} = \frac{928}{17} = 54.59$

b) The data arranged in ascending order is:

48, 50, 50, 50, 50, 53, 53, 54, 55, 55, 55, 56, 56, 58, 59, 63, 63.

The median is the middle number and its value is 55. The mean is slightly less in value than the median.

3.3 The mean for the sample of 14 values is $\frac{[13(119.7692) + 159]}{14} = \frac{[1557 + 159]}{14} = \frac{1716}{14} = 122.57$

3.5 a) No value in the data set is greater than 4. The mean cannot be larger than the largest observation. Also, since the total must be a whole number and the number of observations is 8, the decimal part of the mean must be 1/8, 2/8, 3/8, ... , or 7/8. These fractions have decimal equivalents of .125, .250, ... , .875 so the mean could not be 4.2.

b) The total would have to be 8 and hence all 8 observations would have to be 1.

c) The sample median would be either a whole number or the average of two whole numbers. Thus the decimal portion of the median would be .0 or .5. Hence 4.2 could not be the median of the 8 data values.

3.7 The ordered values are:

Diet 1: 5, 7, 12 , 13, 15
Diet 2: 3, 4, 11, 13, 14

Since each ordered observation in diet 1 is at least as large as the corresponding ordered observation in diet 2 (with some larger), it follows that the total for diet 1 is larger than the total for diet 2. Since the sample sizes are the same, it then follows that the average weight gain is larger for diet 1 than for diet 2.

3.9 The four observations ordered are: 27.3, 30.5, 33.5, 36.7.

The median = 32 and the mean = 128/4 = 32. If we add the value 32 to the data set, the new data set ordered is 27.3, 30.5, 32, 33.5, 36.7. For this data set, the sample median is 32 and the mean is $\bar{x} = \frac{160}{5} = 32$.

3.11 a) $\frac{7}{10} = .7$

b) $\frac{(1 + 1 + 0 + 1 + 1 + 1 + 0 + 0 + 1 + 1)}{10} = \frac{7}{10}$. They are the same.

c) There need to be 25(.8) = 20 successes in all. Since there are 7 in the first ten, there need to be 13 successes in the 15 added observations.

3.13 The total for the 19 papers is 19(70) = 1330. To have a class average of 71, the total of all 20 papers would have to be 20(71) = 1420. Thus, the last paper would have to be a 1420 – 1330 = 90. To have a class average of 72, the total of all 20 papers would have to be 20(72) = 1440. Thus the last paper would have to be a 1440 – 1330 = 110. (An impossibility, since the maximum score possible is 100).

Section 3.2

3.15 a) There are six negative deviations and two positive deviations. The sum of the eight deviations is zero.

b) $s^2 = \frac{586.875}{7} = 83.8393$, $s = \sqrt{83.8393} = 9.1564$

3.17 a) $\bar{x} = \frac{532}{11} = 48.364$

b)

Observation	Deviation	$(\text{Deviation})^2$
62	13.636	185.9405
23	−25.364	643.3325
27	−21.364	456.4205
56	7.636	58.3085
52	3.636	13.2205
34	−14.364	206.3245
42	−6.364	40.5005
40	−8.364	69.9565
68	19.636	385.5725
45	−3.364	11.3165
83	34.636	1199.6525
	−0.004	3270.5455

$$s^2 = \frac{3270.5455}{10} = 327.05, \quad s = \sqrt{327.05} = 18.08$$

The variance, s^2, could be interpreted as the mean squared deviation from the average distance when detection takes place. This is 327.05 squared centimeters. The standard deviation s could be interpreted as the typical amount by which a distance deviates from the average distance when detection first takes place. This is 18.08 centimeters.

c) Since 20 is 3 units farther below the mean than is 23 and 86 is three units farther above the mean than is 83, the mean of the new data set is the same as the mean of the original data set. However, the deviations for 20 and 86 would be three units greater in absolute value than the deviations for 23 and 83. The other deviations would remain unchanged. Therefore, the variance would increase in value. Hence, the standard deviation would increase in value.

3.19 a) $\Sigma x = (2.75 + 2.62 + 2.74 + \ldots + 3.01) = 56.8$

$\Sigma x^2 = (2.75^2 + 2.62^2 + 2.74^2 + \ldots + 3.01^2) = (7.5625 + 6.8644 + 7.5076 + \ldots + 9.0601) = 197.804$

b)

$$s^2 = \frac{197.804 - \frac{(56.8)^2}{17}}{17-1} = \frac{197.804 - 189.7788}{16} = \frac{8.0252}{16} = .5016$$

$$s = \sqrt{.5016} = .7082$$

s could be interpreted as the typical amount by which an area of scleral lamina deviated from the mean area of scleral lamina.

3.21 a) For the cancer group: $\bar{x} = \dfrac{958}{42} = 22.81$ and the median $= \dfrac{(16 + 16)}{2} = 16$

For the no-cancer group: $\bar{x} = \dfrac{747}{39} = 19.15$ and the median = 12.

The values of both the mean and median suggest that the center of the cancer sample is to the right (larger in value) of the center of the no-cancer group.

b) For the cancer group:

$$s^2 = \frac{62936 - \frac{(958)^2}{42}}{41} = \frac{62936 - 21851.5238}{41} = \frac{41084.4762}{41} = 1002.0604$$

$$s = \sqrt{1002.0604} = 31.6553$$

For the no-cancer group:

$$s^2 = \frac{25277 - \frac{(747)^2}{39}}{38} = \frac{25277 - 14307.9231}{38} = \frac{10969.0769}{38} = 288.6599$$

$$s = \sqrt{288.6599} = 16.99$$

The values of s^2 and s suggest that there is more variability in the cancer sample values than in the no-cancer sample values.

For the cancer group, the iqr = 22 – 11 = 11

For the no-cancer group, the iqr $= \dfrac{29 + 24}{2} - \dfrac{8 + 9}{2} = 26.5 - 8.5 = 18$

The previous conclusion about variability is not confirmed by the interquartile ranges. A possible explanation is the extreme outlier in the cancer sample whose value is 210. Eliminating the largest value in each group and recalculating the variances yield the following results.

For the cancer group: $n = 41$, $\Sigma x = 748$, $\Sigma x^2 = 18836$

$$s^2 = \frac{18836 - \frac{(748)^2}{41}}{40} = \frac{18836 - 13646.439}{40} = \frac{5189.561}{40} = 129.739$$

$$s = \sqrt{129.739} = 11.39$$

For the no-cancer group: $n = 38 \quad \Sigma x = 662, \quad \Sigma x^2 = 18052$

$$s^2 = \frac{18052 - \frac{(662)^2}{38}}{37} = \frac{18052 - 11532.7368}{37} = \frac{6519.2632}{37} = 176.1963$$

$$s = \sqrt{176.1963} = 13.27$$

The values of s for the two groups with the outliers removed are closer in value and the no-cancer group is higher.

3.23 Multiplying each data point by 10 yields

x	$x - \bar{x}$	$(x - \bar{x})^2$
620	136.36364	18595.04132
230	−253.63636	64331.40496
270	−213.63636	45640.49587
560	76.36364	5831.40496
520	36.36364	1322.31405
340	−143.63636	20631.40496
420	−63.63636	4049.58678
400	−83.63636	6995.04132
680	196.36364	38558.67769
450	−33.63636	1131.40496
830	346.36364	119967.76860
	0.00004	327054.54545

$\bar{x} = 483.63636$

$$s^2 = \frac{(327054.54545)}{10} = 32705.4545 \qquad s = \sqrt{32705.4545} = 180.846$$

The standard deviation for the new data set is 10 times larger than the standard deviation for the original data set.

3.25 The ordered sample is:

lower half: 2.34, 2.43, 2.62, 2.74, 2.74, 2.75, 2.78, 3.01, 3.46

upper half: 3.46, 3.56, 3.65, 3.85, 3.88, 3.93, 4.21, 4.33, 4.52

a) lower quartile = 2.74 upper quartile = 3.88

b) iqr = 3.88 − 2.74 = 1.14

c) The value of the iqr would not be affected by this change since both 5.33 and 5.52 are larger than the upper quartile.

d) The observation 2.34 could be increased to as much as 2.74 (the value of the lower quartile) without affecting the iqr.

e) If an 18th observation, 4.60, is added to the data, then the lower quartile = 2.74, the upper quartile = 3.93 and the iqr = 3.93 – 2.74 = 1.19.

3.27 a) For sample 1, $\bar{x}$ = 7.81 and s = .39847
For sample 2 , $\bar{x}$ = 49.68 and s = 1.73897

b) For sample 1, CV = (100)(.39847)/7.81 = 5.10
For sample 2, CV = (100)(1.73897)/49.68 = 3.50

The two CV values indicate that there is more variability in the weights of cans of 8 ounce pet food than there is variability in 50 pound bags of dry pet food when the variability is expressed as a percent of the mean.

Section 3.3

3.29 a) The value 57 is one standard deviation above the mean. The value 27 is one standard deviation below the mean. By the empirical rule, roughly 68% of the vehicle speeds were between 27 and 57.

b) From part (a) it is determined that 100% – 68% = 32% were either less than 27 or greater than 57. Because the normal curve is symmetric, this allows us to conclude that half of the 32% (which is 16%) falls above 57. Therefore, an estimate of the percentage of fatal automobile accidents that occurred at speeds over 57 mph is 16%.

3.31 Since the histogram is well approximated by a normal curve, the empirical rule will be used to obtain answers for part a) - c).

a) Because 2500 is 1 standard deviation below the mean and 3500 is 1 standard deviation above the mean, about 68% of the sample observations are between 2500 and 3500.

b) Since both 2000 and 4000 are 2 standard deviations from the mean, approximately 95% of the observations are between 2000 and 4000. Therefore about 5% are outside the interval from 2000 to 4000.

c) Since 95% of the observations are between 2000 and 4000 and about 68% are between 2500 and 3500, there is about 95–68 = 27% between 2000 and 2500 or 3500 and 4000. Half of those, 27/2 = 13.5%, would be in the region from 2000 to 2500.

d) When applied to a normal curve, Chebyshev's rule is quite conservative. That is, the percentages in various regions of the normal curve are quite a bit larger than the values given by Chebyshev's rule.

3.33 a) median = 74, upper quartile = 80, lower quartile = 65

b) iqr = 80 – 65 = 15
lower inner fence = 65 – 1.5(15) = 65 – 22.5 = 42.5
lower outer fence = 65 – 3(15) = 65 – 45 = 20
40 is a mild outlier because it is between the lower inner fence (42.5) and lower outer fence (20).

c) 15 is an extreme outlier because it is less than the lower outer fence.

d) The upper inner fence has value 80 + 1.5(15) = 102.5. Since this was a 100 point test, it would be impossible to have an outlier or outliers on the high end.

3.35 For the first test the student's z-score is $\frac{(625 - 475)}{100} = 1.5$ and for the second test it is $\frac{(45 - 30)}{8} = 1.875$. Since the student's z-score is larger for the second test than for the first test, the student's performance was better on the second exam.

3.37 a)

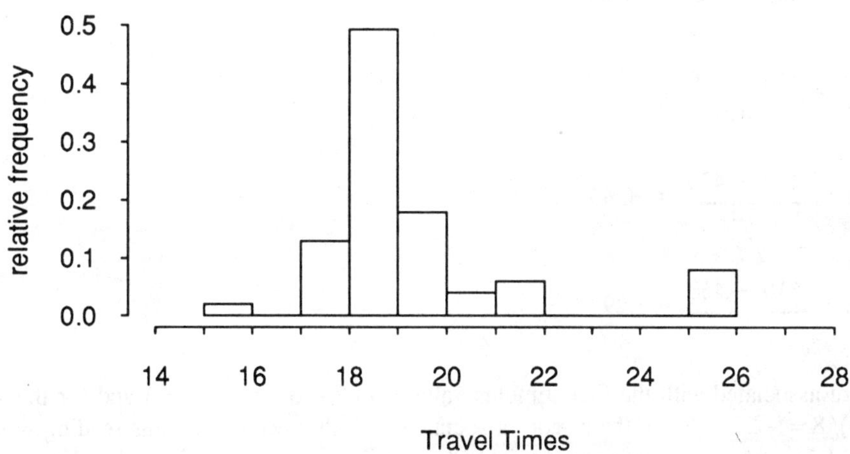

b) The 15th percentile is approximately 18, because the cumulative relative frequency for the 17 -< 18 class is .15.

The 86th percentile is approximately 21, because the cumulative relative frequency for the 20 -< 21 class is .86.

c) The 90th percentile is approximately 21.67, because the cumulative relative frequency up to 21 is 86%, up to 22 is 92%, and 21.67 is two-thirds of the way across the class of 21 -< 22. The value two-thirds was chosen because 90% is two-thirds of the way between 86% and 92%.

d) The 10th percentile will be $\frac{(10 - 2)}{13} = \frac{8}{13}$ of the way across the 17 -< 18 class. Hence, the value of the 10th percentile is 17.615.

3.39 a) A z-score of +2.2 indicates that this person scored 2.2 standard deviations above the mean. Roughly 95% of the scores are within 2 standard deviations of the mean, so this person scored in the upper 2.5% of the class.

b) A z score of −.4 indicates that this person scored .4 standard deviations below the mean.

c) A z score of −1.8 indicates that this person scored 1.8 standard deviations below the mean.

d) A z score of 1.0 indicates that this person scored 1 standard deviation above the mean. Roughly 68% of the scores are within 1 standard deviation of the mean. This means that about (1−.68)/2 = 16% scored higher than the person. Another way of stating this is that this person scored at the 84th percentile.

e) A z score of 0 indicates that this person scored at the mean. Because a normal curve is symmetric, the mean is also the median. Thus half of the scores were higher than this person's score and half were lower. Another way of saying this, is that this person scored at the 50th percentile.

3.41 a) $$z = \frac{320 - 450}{70} = -1.86$$

b) $$z = \frac{475 - 450}{70} = 0.36$$

c) $$z = \frac{420 - 450}{70} = -0.43$$

d) $$z = \frac{610 - 450}{70} = 2.29$$

3.43 The z-score associated with the first stimulus value is (4.2 − 6.0)/1.2 = −1.5 and for the second stimulus value it is (1.8 − 3.6)/.8 = −2.25. Since the z-score associated with the second stimulus reading is more negative (−2.25 is less than −1.5), you are reacting more quickly (relatively) to the second stimulus than to the first.

Supplementary Exercises

3.45 The data arranged in ascending order is:

Iron	Zinc
1.5	39
1.8	50
2.5	62
2.9	66
3.2	66
3.3	67
3.4	68
3.4	69
3.9	89
4.0	110
4.5	220

a) For the iron data: $\bar{x}$ = 34.4/11 = 3.1273 and the median is 3.3. Yes, the numerical values of the mean and median are roughly equal for the iron data.

b) For the zinc data: $\bar{x}$ = 906/11 = 82.364 and the median is 67. No, the numerical value of the mean is quite a bit larger than the median for the zinc data. The median would best represent the typical center value because of the two very large positive values 110 and 220.

c) The zinc data would have the larger variance because there is a lot more variability in the zinc data than in the iron data. This is suggested by the fact that the zinc data ranges from 39 to 220, while the data for Iron ranges from 1.5 to 4.5.

3.47 The z score for 200 is (200 – 125)/75 = 1 and the z score for 275 is (275 – 125)/75 = 2. Without assuming anything about the shape of the distribution, nothing can be said about the proportion of days on which a first stage smog alert is declared. However, since on at least 75% of the days the smog index is less than 275, we can conclude that a second stage smog alert is declared on, at most, 25% of the days.

3.49 $n = 20,\ \Sigma x = 18.51,\ \Sigma x^2 = 17.2555$

$$\bar{x} = \frac{18.51}{20} = .9255$$

$$s^2 = \frac{17.2555 - \frac{(18.51)^2}{20}}{19} = \frac{17.2555 - 17.1310}{19} = \frac{.1245}{19} = .0066$$

$$s = \sqrt{.0066} = .0809$$

The data arranged in ascending order is:

.78, .81, .81, .85, .85, .86, .92, .92, .93, .93, .93, .93, .95, .95, .96, .96, 1.00, 1.05, 1.06, 1.06

The median value is (.93 + .93)/2 = .93

The boxplot for the cadence data is shown below.

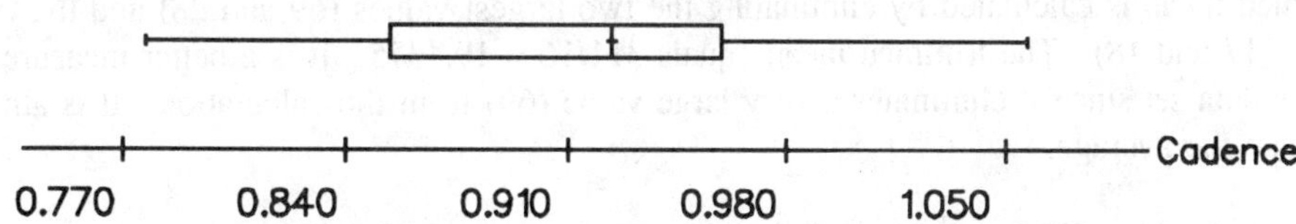

The typical cadence is about .93 strides per second. There is not a large amount of variability in strides per second since the standard deviation is .0809. That is, the variability is small relative to the mean value. The boxplot reveals a bit of skewness in the middle fifty percent of the data. However, overall the distribution of cadence is reasonably symmetric.

3.51

x	$x - \bar{x}$	$(x - \bar{x})^2$
18	−4.15	17.22
18	−4.15	17.22
25	2.85	8.12
19	−3.15	9.92
23	0.85	0.72
20	−2.15	4.62
69	46.85	2194.92
18	−4.15	17.22
21	−1.15	1.32
18	−4.15	17.22
18	−4.15	17.22
20	−2.15	4.62
18	−4.15	17.22
18	−4.15	17.22
20	−2.15	4.62
18	−4.15	17.22
19	−3.15	9.92
28	5.85	34.22
17	−5.15	26.52
18	−4.15	17.22
443	0.00	2454.48

a) $\bar{x} = \dfrac{443}{20} = 22.15$

$s^2 = 2454.48/19 = 129.183$

$s = \sqrt{129.183} = 11.366$

b) The 10% trimmed mean is calculated by eliminating the two largest values (69 and 28) and the two smallest values (17 and 18). The trimmed mean equals 311/16 = 19.4375. It is a better measure of location for this data set since it eliminates a very large value (69) from the calculation. It is almost 3 units smaller than the average.

c) The upper quartile is (21 + 20)/2 = 20.5, the lower quartile is (18 + 18)/2 = 18, and the iqr = 20.5 − 18 = 2.5.

d) upper quartile + 1.5(iqr) = 20.5 + 1.5(2.5) = 24.25
upper quartile + 3.0(iqr) = 20.5 + 3(2.5) = 28.00
The values 25 and 28 are mild outliers and 69 is an extreme outlier.

e)

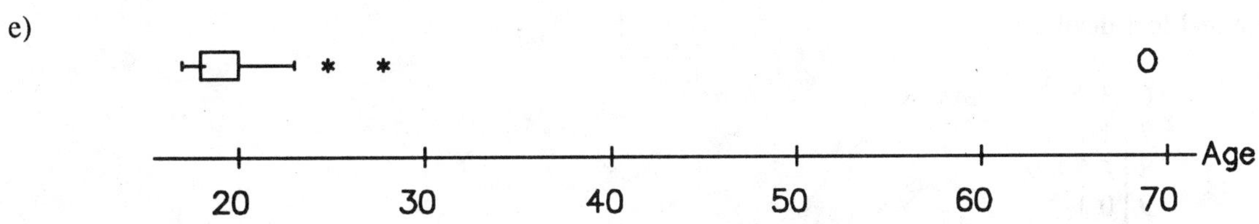

3.53 a) If one observation is deleted at each end, the trimming percentage would be 1/15 = .0667 or 6.67%. If two observations are deleted at each end, the trimming percentage would be 2/15 = .1333 or 13.33%.

b) The ordered data is 7.9, 8.5, 8.8, 9.2, 9.3, 9.6, 9.8, 10.5, 10.7, 11.0, 12.1, 12.2, 13.2, 13.7, 16.6.
The 6.67% trimmed mean is (8.5 + 8.8 + ... + 13.7)/13 = 138.6/13 = 10.66.
The 13.33% trimmed mean is(8.8 + 9.2 + ... + 13.2)/11 = 116.4/11 = 10.58

c) To calculate a 10% trimmed mean, one might interpolate linearly between the two computed trimmed means in b). The resulting value would be

$$\frac{10.58 - 10.66}{.1333 - .0667} = \frac{x - 10.66}{.10 - .0667} \Rightarrow \frac{-.08}{.0667} = \frac{x - 10.66}{.0333}$$

$$x = 10.66 + .0333\left(\frac{-.08}{.0667}\right) = 10.66 + .5(-.08) = 10.62$$

Or, since 10% is halfway between 6.67% and 13.33%, one could just average the two trimmed means already calculated.

3.55 $\Sigma x = 268.8$ $\Sigma x^2 = 2756.54$

$$\bar{x} = \frac{268.8}{27} = 9.956$$

Median value = 10.6

$$s^2 = \frac{2756.54 - \frac{(268.8)^2}{27}}{26} = \frac{2756.54 - 2676.053}{26} = \frac{80.487}{26} = 3.096$$

$$s = \sqrt{3.096} = 1.759$$

A stem and leaf display is

Stem	Leaf
6	3 4
7	1 7
8	4 5 8 9
9	0 1
10	0 1 2 6 6 7 7 8 9
11	1 2 2 4 9 9
12	2
13	1

stem: ones
leaf: tenths

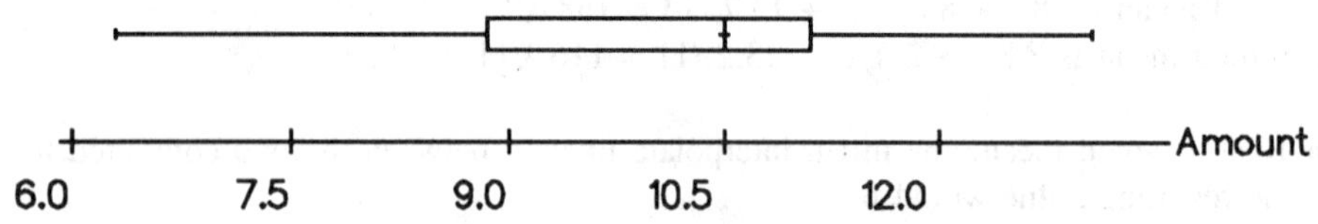

3.57 a) The sample median and $\bar{x}$ will be identical if, for every value below the mean, there is a corresponding value above the mean, such that these two values differ from the mean by the same amount.

Also, when the average with the sample median excluded from the data is the same as the average with the sample median included.

b) The trimmed mean and $\bar{x}$ will be identical when the average of the two deleted values equals the value of the trimmed mean. Or, when the two deleted values differ from the mean by the same value.

3.59 $$\bar{x}_w = \frac{120000(268) + 30000(210) + 45000(220)}{120000 + 30000 + 45000} = \$248$$

3.61 a) The 16th and 84th percentiles are the same distance from the mean but on opposite sides. Since 80 is 20 units below 100, the 84th percentile is 20 units above 100, which would be at 120.

b) Since 84 – 16 = 68, roughly 68% of the scores are between 80 and 120. Thus, by the empirical rule, 120 is one standard deviation above the mean, so the standard deviation has value 20.

c) The value 90 would have a z-score of (90 – 100)/20 = –1/2.

d) The value 140 would be two standard deviations above the mean and thus roughly .5(5%) = 2.5% of the scores would be larger than 140. Hence 140 would be the 97.5 percentile.

e) The value 40 is three standard deviations below the mean, so only about .5(.3%) or .15% of the scores are below 40. There would not be many scores below 40.

Chapter 4
Probability

Section 4.1

4.1 a) Sample Space = {AA, AM, MA, MM}

b)

First Car	Second Car	Outcome
A	A	A A
	M	A M
M	A	M A
	M	M M

c) B = {MA, AM, AA}; C = {AM, MA}; D = {MM} D is a simple event.

d) (B and C) = {AM, MA}; (B or C) = {MA, AM, AA}

4.3 a)

Headlights	Tires	Outcome
0	0	0 0
	1	0 1
	2	0 2
	3	0 3
	4	0 4
1	0	1 0
	1	1 1
	2	1 2
	3	1 3
	4	1 4
2	0	2 0
	1	2 1
	2	2 2
	3	2 3
	4	2 4

b) not A = {(2,0), (2,1), (2,2), (2,3), (2,4)}

A or B = {(0,0), (0,1), (0,2), (0,3), (0,4), (1,0), (1,1), (1,2), (1,3), (1,4), (2,0), (2,1)}

A and B = {(0,0), (0,1), (1,0), (1,1)}

c) A and C are not disjoint because A and C = {(0,4), (1,4)}, which is not the empty set.

B and C are disjoint because (B and C) is the empty set. (A car cannot have 4 defective tires and at the same time have at most 1 defective tire).

4.5 a)

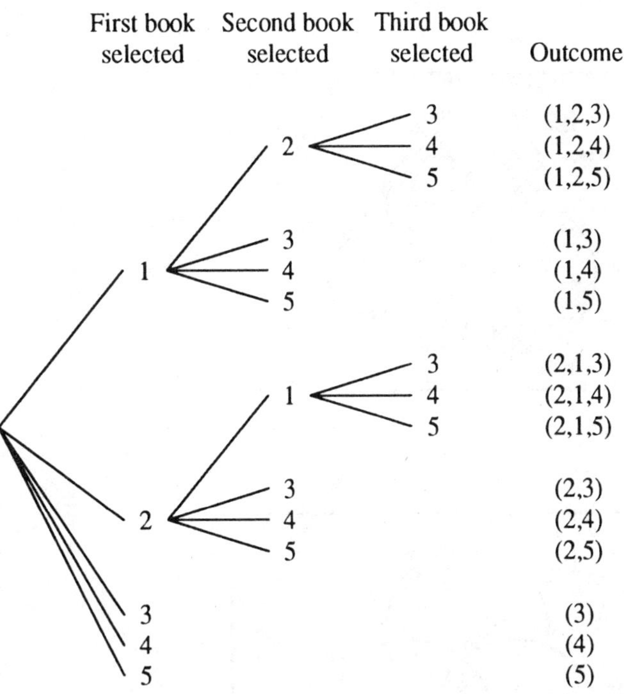

b) A = {(3) , (4) , (5)}

c) C = {(1,2,5) , (1,5) , (2,1,5) , (2,5) , (5)}

4.7 a) A = {NN, DNN, NDN}

b) B = {DDNN, NDDN, DNDN}

c) There are countably infinitely many.

4.9 a)

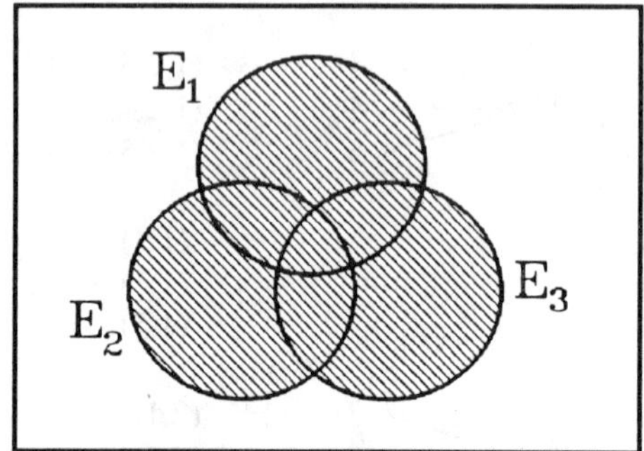

b)

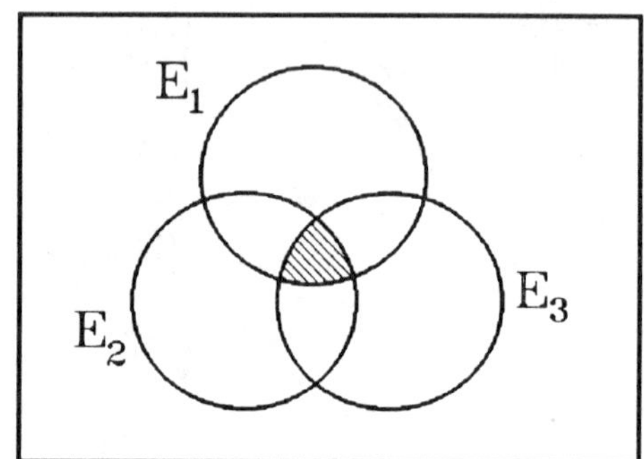

c)

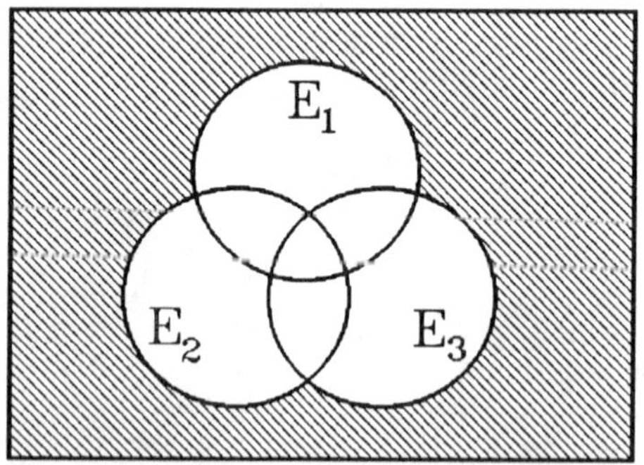

d)

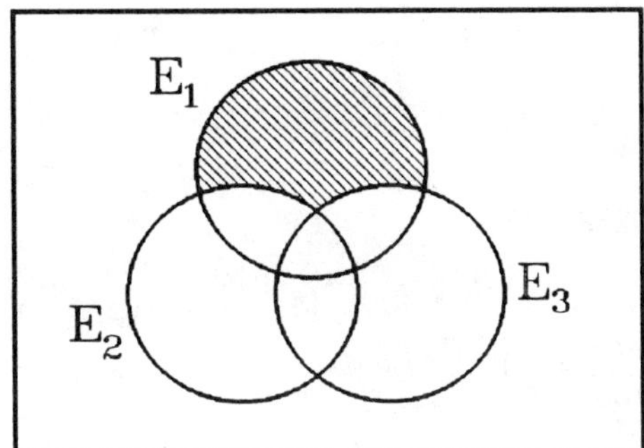

e)

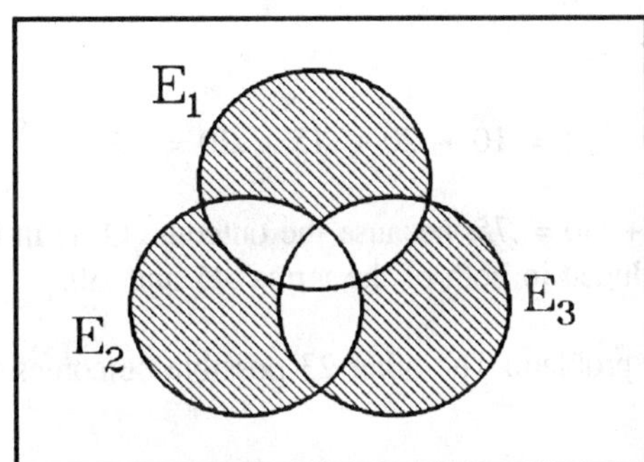

f)

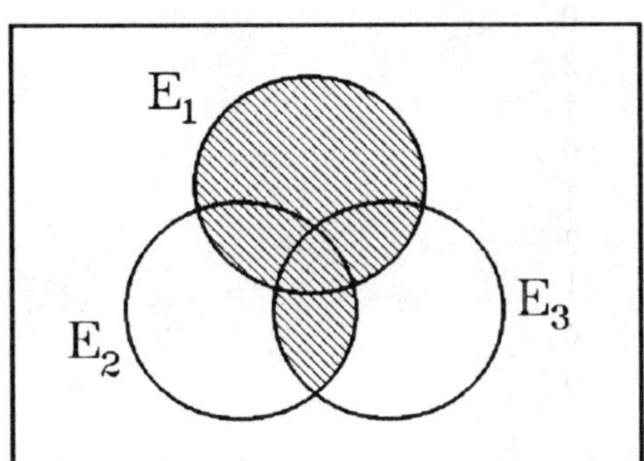

Section 2

4.11 a) P(Dreyer's ice cream is purchased) = .20 + .25 = .45

b) P(Von's brand is not purchased) = .10 + .15 + .20 + .25 = .70

c) P(size purchased is larger than a pint) = .20 + .25 + .30 = .75

4.13 a) $P(\text{twins}) = \dfrac{500{,}000}{42{,}005{,}100} = .0119$

b) $P(\text{quadruplets}) = \dfrac{100}{42{,}005{,}100} = .00000238$

c) $P(\text{more than a single child}) = \dfrac{505{,}100}{42{,}005{,}100} = .012$

4.15 a) 7%

b) 15% + 10% + 5% = 30%

c) 1 − [.18 + .25] = 1 − .43 = .57 or 57%

4.17 P(midsize or 4.375" grip) = P(O_1 or O_2 or O_3 or O_4) = .10 + .20 + .15 + .20 = .65

So P(midsize or 4.375" grip) is not equal to .45 + .30 = .75, because the outcome O_1 is in both of the events midsize and 4.375" grip. Hence the P(O_1) is included in both of the terms .45 and .30.

4.19 a) The outcome (2, 4, 3, 1) is given in the problem. The other 23 possible outcomes with the number of correct owners listed alongside are:

(1, 2, 3, 4)	4	(3, 1, 2, 4)	1
(1, 2, 4, 3)	2	(3, 1, 4, 2)	0
(1, 4, 2, 3)	1	(3, 4, 1, 2)	0
(1, 4, 3, 2)	2	(3, 4, 2, 1)	0
(1, 3, 4, 2)	1	(3, 2, 1, 4)	2
(1, 3, 2, 4)	2	(3, 2, 4, 1)	1
(2, 1, 3, 4)	2	(4, 1, 2, 3)	0
(2, 1, 4, 3)	0	(4, 1, 3, 2)	1
(2, 4, 1, 3)	0	(4, 2, 3, 1)	2
(2, 3, 1, 4)	1	(4, 2, 1, 3)	1
(2, 3, 4, 1)	0	(4, 3, 1, 2)	0
		(4, 3, 2, 1)	0

b) (1, 2, 4, 3), (1, 4, 3, 2), (1, 3, 2, 4), (2, 1, 3, 4), (3, 2, 1, 4), (4, 2, 3, 1)

P(exactly two of the books are returned to their correct owners) = $\dfrac{6}{24} = .2500$

c) P(exactly one receives his/her book) = $\dfrac{8}{24} = .3333$

d) P(exactly three receive their own book) = 0

e) P(at least two of the four students receive their own books) = $\dfrac{(6 + 1)}{24} = \dfrac{7}{24} = .2917$

4.21 a) The ten simple events are (B, C), (B, M), (B, P), (B, S), (C, M), (C, P), (C, S), (M, P), (M, S), (P, S).

b) Each would be equally likely with probability 0.1.

c) P(committee includes the statistics representative) = P[(B, S), (C, S), (M, S), (P, S)] = .4

d) P(both members are from "laboratory resources") = P[(B, C), (B, P), (C, P)] = .3

4.23 The simple events are (1, 2), (1, 3), (1, 4), (1, 5), (2, 3), (2, 4), (2, 5), (3, 4), (3, 5), (4, 5).

a) P(both are statistics majors) = P[(1, 2)] = $\frac{1}{10}$

b) P(both are math majors) = P[(3, 4), (3, 5) or (4, 5)] = $\frac{3}{10}$

c) P(at least one is a statistics major) = 1 − P(neither is a statistics major) = 1 − $\frac{3}{20}$ = $\frac{7}{10}$

d) P(students have different majors) = $\frac{6}{10}$

Section 3

4.25 a) $P(E|F) = \frac{P(E \text{ and } F)}{P(F)} = \frac{.54}{.6} = .9$

b) $P(F|E) = \frac{P(E \text{ and } F)}{P(E)} = \frac{.54}{.7} = .771$

c) E and F are not independent because $P(E|F) \neq P(E)$.

4.27 a) $P(E) = \frac{13}{52} = \frac{1}{4}$; $P(F) = \frac{12}{52} = \frac{3}{13}$

b) $$P(E|F) = \frac{P(E \text{ and } F)}{P(F)} = \frac{\left(\frac{3}{52}\right)}{\left(\frac{3}{13}\right)} = \frac{13}{52} = \frac{1}{4}$$

$$P(F|E) = \frac{P(E \text{ and } F)}{P(E)} = \frac{\left(\frac{3}{52}\right)}{\left(\frac{13}{52}\right)} = \frac{3}{13}$$

c) E and F are independent since knowing you have a face card gives no information regarding whether or not the card is a heart.

4.29 The two events are not independent since P(F|E) = 2/3, and this is not equal to P(F) = 1/3.

4.31 A : over 6' B: professional basketball player
P(A|B) represents the proportion of professional basketball players who are over 6'.
P(B|A) represents the proportion of 6' males who are professional basketball players.
Since almost all professional basketball players are over 6', P(A|B) is close to 1. A small percentage of all 6' males are professional basketball players, so P(B|A) is close to 0. Hence, P(A|B) is larger than P(B|A).

4.33 a) $P(A) = .32 + .27 + .18 = .77$

$P(FD) = .27 + .04 = .31$

b) $$P(A|FD) = \frac{P(A \text{ and } FD)}{P(FD)} = \frac{.27}{.31} = .871$$

c) $$P(M|\text{not } HB) = \frac{P(M \text{ and not } HB)}{P(\text{not } HB)} = \frac{(.08 + .04)}{(.40 + .31)} = \frac{.12}{.71} = .169$$

The conditional probability of manual transmission given not a hatchback is smaller than the unconditional probability of a manual transmission.

4.35 a) $$P(E) = \frac{6}{10}$$

b) $$P(F|E) = \frac{5}{9}$$

c) $$P(E \text{ and } F) = P(E)P(F|E) = \left(\frac{6}{10}\right)\left(\frac{5}{9}\right) = \frac{30}{90} = \frac{1}{3}$$

4.37 a) The expert assumed that the positions of the two valves were independent.

b) If the car were driven in a straight line, the relative positions of the valves would remain unchanged. Thus, the probability of them ending up at a one o'clock, six o'clock position would be 1/12, not 1/144. The value 1/144 is smaller than the correct probability of occurrence.

4.39 a) Yes, E_1 and E_2 are dependent events because the chance of a defective board on the second selection depends upon what occurred on the first selection.

b) $$P(\text{not } E_1) = \frac{4960}{5000} = .992$$

c) $$P(E_2|E_1) = \frac{39}{4999} = .0078$$

$$P(E_2|\text{not } E_1) = \frac{40}{4999} = .0080$$

They are close to being equal, but they are not exactly the same.

d) It would be reasonable to view E_1 and E_2 as being independent for all practical purposes, since the two probabilities in part (c) are so close to being equal in value.

4.41 $P(E_1|E_2) = 0$ since E_1 and E_2 are disjoint.

$P(E_2|E_1) = 0$ since E_1 and E_2 are disjoint.

Disjoint events are not independent, since $P(E_1|E_2) = 0$ which does not equal $P(E_1)$ if $P(E_1) \neq 0$.

Section 4

4.43 a) P(E or F) = P(E) + P(F) − P(E and F) = .4 + .3 − .15 = .55

b) P(neither E nor F) = 1 − P(E or F) = 1 − .55 = .45

c) P(exactly one) = P(E or F) − P(E and F) = .55 − .15 = .40

d) P(must stop at first light only) = P(E) − P(E and F) = .4 − .15 = .25

4.45 a) P(medium auto and high homeowner) = .10

b) P(low auto) = .04 + .06 + .05 + .03 = .18

P(low homeowner) = .06 + .10 + .03 = .19

c) P(in same category for both types of insurance) = P(L, L) + P(M, M) + P(H, H) = .06 + .20 + .15 = .41

d) P(not in same category for both types)
= 1 − P(in same category for both types) = 1 − .41 = .59

e) P(an individual has at least one low deductible)
= P(low auto) + P(low homeowner) − P(both low)
= .18 + .19 − .06 = .31

f) P(neither deductible level is low) = 1 − P(at least one is low) = 1 − .31 = .69

4.47 a)

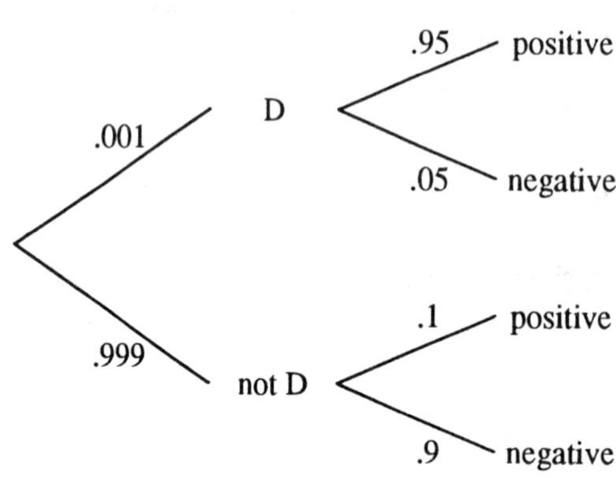

b) P(has disease and positive result) = .001(.95) = .00095

c) P(positive result) = P(has disease and positive) + P(not disease and positive)
= .001(.95) + .999(.1) = .00095 + .0999 = .10085

d) P(has disease | positive result) = $\dfrac{.00095}{.10085}$ = .00942.

The 95% rate for testing positive when a person has the disease makes the test appear to be very accurate. The computed conditional probability of .00942 is very small. Because the disease is rare, most positive results come about from errors rather than for persons who have the disease.

4.49 $P(E_1 \text{ and } L) = P(E_1)P(L|E_1) = .4(.02) = .008$

Section 5

4.51 There are 27(41) = 1,107 ways to listen first to a Mozart piano concerto and then to a symphony.

4.53 There are 6(5)(3)(8) = 720 ways to select a single contractor of each type.

4.55 a) 6(5)(4) = 120

b) $\binom{15}{3} = \dfrac{15!}{3!12!} = \dfrac{15(14)(13)}{3(2)(1)} = \dfrac{2730}{6} = 455$

c) $\dfrac{120}{455} = 0.2637$

4.57 5(8)(6) = 240 nights.

4.59 a) $P_{15,9} = 15(14)(13)(12)(11)(10)(9)(8)(7) = 1{,}816{,}214{,}400$

b) If A, B, C and D are left on the bench, then the starting nine must be selected from the remaining eleven players. This can be done in $P_{11,9} = 11(10)(9)(8)(7)(6)(5)(4)(3) = 19{,}958{,}400$ ways. Hence the probability that players A, B, C and D will be left on the bench is

$$\frac{19{,}958{,}400}{1{,}816{,}214{,}400} = 0.011.$$

Supplementary Exercises

4.61 P(at least two disks must be examined) = 1 − P(only one examined) = 1 − P(first examined disk is blank)

$$= 1 - \frac{15}{25} = \frac{10}{25} = .4$$

4.63 The possible peer review committees are

(3, 6) (3,7) (3, 10) (3, 14) (6, 7) (6, 10) (6, 14) (7, 10) (7, 14) (10, 14).

The underlined committees have a total of at least 15 years experience. The probability of selecting a committee with representatives that have a total of at least 15 years of teaching experience is 6/10 = .6.

4.65 P(a randomly selected subscriber reads at least one of these three columns)
= .14 + .23 + .37 − .08 − .09 −.13 + .05 = 0.49

4.67 R_1 and R_2 are not independent events since $P(R_2|R_1) = 1139/2516 = .4527$ which is not equal to $P(R_2)$ which is .4529. However, these values do not differ much and thus one could say they are independent for practical purposes.

4.69 a) $$P(\text{medium}|\text{short-sleeve plaid}) = \frac{P(\text{medium and short-sleeve plaid})}{P(\text{short-sleeve plaid})} = \frac{.08}{(.04 + .08 + .03)} = \frac{.08}{.15} = .533$$

b) $$P(\text{medium}|\text{short-sleeve plaid}) = \frac{P(\text{short-sleeve and medium plaid})}{P(\text{medium plaid})} = \frac{.08}{(.08 + .10)} = \frac{.08}{.18} = .444$$

$$P(\text{long-sleeve}|\text{medium plaid}) = \frac{P(\text{long-sleeve and medium plaid})}{P(\text{medium plaid})} = \frac{.10}{(.08 + .10)} = \frac{.10}{.18} = .556$$

4.71 a) Yes, if the slip that reads "win prizes 1, 2 and 3" is drawn. $P(E_1 \text{ and } E_2) = 1/4$, $P(E_1) = 2/4$, and $P(E_2) = 2/4$. Since $P(E_1 \text{ and } E_2) = P(E_1)P(E_2)$, the events E_1 and E_2 are independent.

b) $P(E_1 \text{ and } E_3) = 1/4$, $P(E_1) = 1/2$, and $P(F_3) = 1/2$
Since $P(E_1 \text{ and } E_3) = P(E_1)P(E_3)$, the events E_1 and E_3 are independent.

$P(E_2 \text{ and } E_3) = 1/4$, $P(E_2) = 1/2$, and $P(E_3) = 1/2$
Since $P(E_2 \text{ and } E_3) = P(E_2)P(E_3)$, the events E_2 and E_3 are independent.

c) $P(E_1 \text{ and } E_2 \text{ and } E_3) = \frac{1}{4}$

$P(E_1)(E_2)(E_3) = \frac{1}{2}\left(\frac{1}{2}\right)\left(\frac{1}{2}\right) = \frac{1}{8}$

So $P(E_1 \text{ and } E_2 \text{ and } E_3) \neq P(E_1)P(E_2)P(E_3)$

4.73 a) P(fills and pays with a credit card) = P(fills)P(pays with credit card | fills) = .4(.8) = .32

b) P(all three fill their tanks and pay by credit card) = $(.32)^3$ = .032768

4.75 a) $P(A_1) = \frac{3}{15} = .2$

$P(A_2|A_1) = \frac{2}{14} = .143$

$P(A_1 \text{ and } A_2)$ = P(both applications selected for processing are for 36 months) = $P(A_1)P(A_2|A_1)$

$= \left(\frac{3}{15}\right)\left(\frac{2}{14}\right) = \frac{1}{35} = .0286$

b) P(both are the same duration)
= P(both are for 36 months) + P(both are for 48 months) + P(both are for 60 months)

$= \left(\frac{3}{15}\right)\left(\frac{2}{14}\right) + \left(\frac{5}{15}\right)\left(\frac{4}{14}\right) + \left(\frac{7}{15}\right)\left(\frac{6}{14}\right)$

$= \frac{(6 + 20 + 42)}{210} = \frac{68}{210} = .324$

4.77 a) $P(B_1 \text{ and } S) = P(S)P(B_1|S) = .2(.5) = .10$

$P(T) = 1 - P(S) = .8$

$P(B_2 \text{ and } T) = P(T) - P(B_1 \text{ and } T) - P(B_3 \text{ and } T) = .8 - .35 - .26 = .19$

$P(B_2 \text{ and } T) = P(B_2)P(T|B_2) \Rightarrow .19 = P(B_2)(.76)$

$P(B_2) = \frac{.19}{.76} = .25$

$P(B_2 \text{ and } S) = P(B_2) - P(B_2 \text{ and } T) = .25 - .19 = .06$

$P(B_3 \text{ and } S) = P(S) - P(B_1 \text{ and } S) - P(B_2 \text{ and } S) = .2 - .10 - .06 = .04$

b) $P(\text{not } B_1|T) = 1 - P(B_1|T) = 1 - \frac{P(B_1 \text{ and } T)}{P(T)} = 1 - \frac{.35}{.80} = \frac{.45}{.80} = .5625$

4.79 P(all four are good) = P(E and F and G and H) = P(E) P(F|E) P[G|(E and F)] P[H|(E and F and G)]

$$= \frac{20}{25}\,\frac{19}{24}\,\frac{18}{23}\,\frac{17}{22} = .8(.7917)(.7826)(.7727) = .3830$$

P(at least one is bad) = 1 − P(all four are good) = 1 − .3830 = .6170

4.81 Let S denote the bit received is passed on with no reversal and F denote the bit received is passed on with a reversal.

a) $P(S_1 \text{ and } S_2 \text{ and } S_3) = P(S_1)(S_2)(S_3) = .8(.8)(.8) = .512$

b) P(1 received if a 1 is sent)
= P(no reversals) + P(2 reversals)
= $P(S_1 \text{ and } S_2 \text{ and } S_3) + P(S_1 \text{ and } F_2 \text{ and } F_3) + P(F_1 \text{ and } F_2 \text{ and } S_3) + P(F_1 \text{ and } S_2 \text{ and } F_2)$
= .8(.8)(.8) + (.8)(.2)(.2) + (.2)(.2)(.8) + (.2)(.8)(.2)
= .512 + 3(.032) = .512 + .096
= .608

4.83 P(all five are of the same brand) = P(all five are Penn) + P(all five are Wilson)

$$= \frac{\binom{12}{5} + \binom{8}{5}}{\binom{20}{5}} = \frac{792 + 56}{15{,}504} = \frac{848}{15{,}504} = 0.0547$$

4.85 P(B) = P(A) + P[(not A) and B]. This says that P(B) must be at least as large as P(A).

4.87 a) P(1-2 subsystem works) = .9(.9) = .81
P(3-4 subsystem works) = .9(.9) = .81

b) P(both subsystems function) = .81(.81) = .6561

c) P(system functions) = P(1-2 subsystem works) + P(3-4 subsystem works) − P(both subsystems function)
= .81 + .81 − .6561 = .9639

Chapter 5
Random Variables and Discrete Probability Distributions

Section 1

5.1 a) discrete

b) continuous

c) discrete

d) discrete

e) continuous

5.3 Possible y values are the positive integers.

Outcomes	**y-value**
S	1
LLS	3
LRS	3
LS	2
RS	2

5.5 y is a continuous variable with values from 0 to 100 ft.

5.7

Outcomes	x-values	y-values	z-values	w-values
(1,2)	3	−1	1	0
(1,3)	4	−2	0	0
(1,4)	5	−3	1	1
(2,3)	5	−1	1	0
(2,4)	6	−2	2	1
(3,4)	7	−1	1	1

Section 2

5.9 a) $p(4) = 1 - [.65 + .20 + .10 + .04] = 1 - .99 = .01$

b) About 20% of the cartons have exactly 1 broken egg.

c) $P(y \leq 2) = p(0) + p(1) + p(2) = .65 + .20 + .10 = .95$
About 95% of the cartons have two or fewer broken eggs.

d) $P(y < 2) = p(0) + p(1) = .65 + .20 = .85$
$P(y < 2)$ is smaller than $P(y \leq 2)$ because $P(y < 2)$ does not include $P(y = 2)$, while $P(y \leq 2)$ does include $P(y = 2)$

e) P(10 unbroken eggs) + P(exactly 2 unbroken eggs) = $P(x = 2) = .10$

f) P(at least 10 unbroken) + P(2 or fewer broken) = $P(y \leq 2) = .95$

5.11 a) $P(x \leq 100) = .05 + .10 + .12 + .14 + .24 + .17 = .82$

b) $P(x > 100) = 1 - P(x \leq 100) = 1 - .82 = .18$

c) $P(x \leq 99) = .05 + .10 + .12 + .14 + .24 = .65$
$P(x \leq 97) = .05 + .10 + .12 = .27$

5.13 The results of the 50 observations on x are:

0,0,1,1,1,1,1,2,1,1,1,1,0,1,0,1,0,2,1,1,0,1,1,1,1,1,1,0,1,1,0,1,1,1,0,1,1,1,1,2,0,1,0,1,0,0,1,1,1,0

Number of Defects	Frequency	Rel. Freq.
0	14	.28
1	33	.66
2	3	.06
		1.00

5.15 a)

Outcomes	Values of x	Probability
SSSS	4	(.2)(.2)(.2)(.2) = .0016
SSSF	3	(.2)(.2)(.2)(.8) = .0064
SSFS	3	(.2)(.2)(.8)(.2) = .0064
SFSS	3	(.2)(.8)(.2)(.2) = .0064
FSSS	3	(.8)(.2)(.2)(.2) = .0064
SSFF	2	(.2)(.2)(.8)(.8) = .0256
SFSF	2	(.2)(.8)(.2)(.8) = .0256
FSSF	2	(.8)(.2)(.2)(.8) = .0256
SFFS	2	(.2)(.8)(.8)(.2) = .0256
FSFS	2	(.8)(.2)(.8)(.2) = .0256
FFSS	2	(.8)(.8)(.2)(.2) = .0256
SFFF	1	(.2)(.8)(.8)(.8) = .1024
FSFF	1	(.8)(.2)(.8)(.8) = .1024
FFSF	1	(.8)(.8)(.2)(.8) = .1024
FFFS	1	(.8)(.8)(.8)(.2) = .1024
FFFF	0	(.8)(.8)(.8)(.8) = .4096

x	0	1	2	3	4
p(x)	.4096	.4096	.1536	.0256	.0016

b) The most likely values for x are 0 and 1. Each has a probability of .4096.

c) $P(x \geq 2) = .1536 + .0256 + .0016 = .180$

5.17 a) The smallest y-value is 1, which results if the experimental outcome S occurs. The second smallest y-value is 2, which results if the experimental outcome FS occurs.

b) The set of y-values is the positive integers.

c)

Outcomes	y-value	Probability
S	1	.7
FS	2	(.3)(.7)
FFS	3	$(.3)^2(.7)$
FFFS	4	$(.3)^3(.7)$
FFFFS	5	$(.3)^4(.7)$

$p(y) = (.3)^{y-1}(.7)$ for y = 1,2,3...

5.19

Outcomes	y-value	Probability
(W,W)	0	(.4)(.4) = .16
(W,T)	1	(.4)(.3) = .12
(W,F)	2	(.4)(.2) = .08
(W,S)	3	(.4)(.1) = .04
(T,W)	1	(.3)(.4) = .12
(T,T)	1	(.3)(.3) = .09
(T,F)	2	(.3)(.2) = .06
(T,S)	3	(.3)(.1) = .03
(F,W)	2	(.2)(.4) = .08
(F,T)	2	(.2)(.3) = .06
(F,F)	2	(.2)(.2) = .04
(F,S)	3	(.2)(.1) = .02
(S,W)	3	(.1)(.4) = .04
(S,T)	3	(.1)(.3) = .03
(S,F)	3	(.1)(.2) = .02
(S,S)	3	(.1)(.1) = .01

y-value	0	1	2	3
p(y)	.16	.33	.32	.19

Section 3

5.21 a) $\mu_y = 0(.65) + 1(.20) + 2(.10) + 3(.04) + 4(.01)$

$\mu_y = 1 + .20 + .20 + .12 + .04 = 0.56$

Some cartons will have no broken eggs, some cartons 1, some cartons 2, some cartons 3, and some cartons 4. The average number of broken eggs per carton is .56.

b) $P(y < \mu_y) = P(y - 0) = .65$

c) $\mu_y \neq \dfrac{(0 + 1 + 2 + 3 + 4)}{5}$ because the number of cartons having 0 broken eggs differs from the number of cartons that have 1 broken egg. Simmilarily for 2, 3, or 4 broken eggs. The y-values must be weighted by the relative frequency of each y-value.

5.23 $\sigma_x^2 = (-1.2)^2(.54) + (-.2)^2(.16) + (.8)^2(.06) + (1.8)^2(.04) + (2.8)^2(.20)$

$= (1.44)(.54) + (.04)(.16) + (.64)(.06) + (3.24)(.04) + (7.84)(.20)$

$= .7776 + .0064 + .0384 + .1296 + 1.568 = 2.52$

$\sigma_x = \sqrt{2.52} = 1.5875$

5.25 a) $\mu = 1(.05) + 2(.10) + 3(.12) + 4(.30) + 5(.30) + 6(.11) + 7(.01) + 8(.01) = 4.12$

b) $\sigma^2 = (1 - 4.12)^2(.05) + (2 - 4.12)^2(.10) + (3 - 4.12)^2(.12) + (4 - 4.12)^2(.30) + (5 - 4.12)^2(.30) + (6 - 4.12)^2(.11) + (7 - 4.12)^2(.01) + (8 - 4.12)^2(.01)$

$= (-3.12)^2(.05) + (-2.12)^2(.10) + (-1.12)^2(.12) + (-.12)^2(.30) + (.88)^2(.30) + (1.88)^2(.11) + (2.88)^2(.01) + (3.88)^2(.01)$

$= .48672 + .44944 + .150528 + .00432 + .23232 + .388784 + .082944 + .150544$
$= 1.9456$

$\sigma = \sqrt{1.9456} = 1.3948$

The variance (1.9456) is the mean squared difference between the number of sets sold and the average number of sets sold. The standard deviation (1.3948) is the typical difference between the number of sets sold and the average number of sets sold.

c) $P(\mu - \sigma \leq x \leq \mu + \sigma) = P(4.12 - 1.3948 \leq x \leq 4.12 + 1.3948)$
$= P(2.7252 \leq x \leq 5.5148) = P(x = 3, 4, \text{ or } 5) = .12 + .30 + .30 = 0.72$

d) $P(x < \mu - 2\sigma \text{ or } \mu + 2\sigma < x) = P(x < 4.12 - 2(1.3948) \text{ or } 4.12 + 2(1.3948) < x)$
$= P(x < 1.3304 \text{ or } 6.9096 < x) = P(x = 1, 7, \text{ or } 8) = .05 + .01 + .01 = 0.07$

5.27 $\mu = 1000(.05) + 5000(.3) + 10000(.4) + 20000(.25) = 10550$

Under the royalty plan the mean amount received would be \$10,550. Since this exceeds the \$10,000 that would be received under the flat payment plan, the author should choose the royalty plan, if the criterion is to maximize the expected renumeration.

5.29 a) y is a discrete random variable since it can take on only six different values.

b) $P(y > 1.20) = .10 + .16 + .08 + .06 = .40$

$P(y < 1.40) = .36 + .24 + .10 + .16 = .86$

c) $\mu = 115.9(.36) + 118.7(.24) + 129.9(.10) + 139.9(.16) + 144.9(.08) + 159.7(.06)$
$= 41.724 + 28.488 + 12.990 + 22.384 + 11.592 + 9.582$
$= 126.760$

$\sigma^2 = (115.9 - 126.760)^2(.36) + (118.7 - 126.760)^2(.24) + (129.9 - 126.760)^2(.10) + (139.9 - 126.760)^2(.16) + (144.9 - 126.760)^2(.08) + (159.7 - 126.760)^2(.06)$

$= 42.4583 + 15.5913 + 0.986 + 27.6255 + 26.3248 + 65.1026$
$= 178.0885$

$\sigma = \sqrt{178.0885} = 13.345$

The mean is the average price per gallon paid by the customers. The standard deviation is the typical difference between the price per gallon paid and the average price per gallon paid.

5.31 a) $y = 100 - 5(x)$

If $x = 1,\; y = 100 - 5 = 95$
$x = 2,\; y = 100 - 10 = 90$
$x = 3,\; y = 100 - 15 = 85$
$x = 4,\; y = 100 - 20 = 80$

b)

y-value	80	85	90	95
p(y)	.1	.3	.4	.2

c) $\mu_y = 80(.1) + 85(.3) + 90(.4) + 95(.2) = 8.0 + 25.5 + 36 + 19 = 88.5$

5.33 $\mu_x^2 = 1^2(.1) + 2^2(.2) + 3^2(.3) + 4^2(.4) = .1 + .8 + 2.7 + 6.4 = 10$

Section 4

5.35 a) $p(2) = P(x = 2) = \binom{4}{2}(.9)^2(.1)^2 = 6(.81)(.01) = .0486$

Only 4.86% of all samples of size 4 would result in exactly 2 households having a VCR.

b) $p(4) = \binom{4}{4}(.9)^4(.1)^0 = 1(.6561)(1) = .6561$

c) $P(x \leq 3) = 1 - P(x = 4) = 1 - .6561 = .3439$

5.37 a) $p(8) = .302$

b) $p(x \leq 7) = p(0) + p(1) + p(2) + p(3) + p(4) + p(5) + p(6) + p(7)$
$= .000 + .000 + .000 + .001 + .006 + .026 + .088 + .201$
$= .322$

or

$P(x \leq 7) = 1 - P(x \geq 8) = 1 - [p(8) + p(9) + p(10)] = 1 - [.302 + .268 + .107] = 1 - .677 = .323$

(The difference of .001 is due to round off error for the tabled values).

c) P(More than half rested or slept) = Px > 5) = P(6) + P(7) + P(8) + P(9) + P(10)
$= .088 + .201 + .302 + .268 + .107 = .966$

5.39 $n = 5,\; \pi = .5$

$P(x = 0) = \frac{5!}{0!5!}(.5)^0(.5)^5 = (.5)^5 = .03125$

$P(x = 1) = \frac{5!}{1!4!}(.5)^1(.5)^4 = 5(.5)^5 = .15625$

$P(x = 2) = \frac{5!}{2!3!}(.5)^2(.5)^3 = 10(.5)^5 = .3125$

$P(x = 3) = \frac{5!}{3!2!}(.5)^3(.5)^2 = 10(.5)^5 = .3125$

$P(x = 4) = \frac{5!}{4!1!}(.5)^4(.5)^1 = 5(.5)^5 = .15625$

$P(x = 5) = \frac{5!}{5!0!}(.5)^5(.5)^0 = (.5)^5 = .03125$

5.41 $n = 10$, $\pi = .5$ (if the graphologist is guessing)

x = number of correct identifications

$P(x \geq 6) = .205 + .117 + .044 + .010 + .001 = .377$

About 38% of the time a person with no expertise would correctly guess six or more times out of ten. This sample does not provide evidence to indicate that the graphologist has an ability to distinquish the handwriting of psychotics.

5.43 $n = 15$ $\pi = .30$ where success is "fail to pass"

a) $P(x \leq 5) = (x = 0) + P(x = 1) + P(x = 2) + P(x = 3) + P(x = 4) + P(x = 5)$
$= .005 + .030 + .092 + .170 + .218 + .207$
$= .722$

b) $P(5 \leq x \leq 10) = P(x = 5) + P(x = 6) + P(x = 7) + P(x = 8) + P(x = 9) + P(x = 10)$
$= .207 + .147 + .081 + .035 + .011 + .003$
$= .484$

c) $\mu = n\pi = 25(.7) = 17.5$,
$\sigma = \sqrt{n\pi(1 - \pi)} = \sqrt{25(.7)(.3)} = \sqrt{5.25} = 2.2913$

d) $P(\mu - \sigma \leq x \leq \mu + \sigma) = P(17.5 - 2.2913 \leq x \leq 17.5 + 2.2913)$
$= P(15.2087 \leq x \leq 19.7913)$
$= P(x = 16, 17, 18, \text{ or } 19)$
$= .134 + .165 + .171 + .148 = 0.618$

5.45 Since the sampling is done without replacement and the sample size is in excess of 5% of the population size, the random variable "number of invalid signatures in a sample of size 1000" would not have a binomial distribution.

5.47 a) $P(x \leq 15)$ when $n = 25$ and $\pi = .8$.
From the binomial tables $P(x \leq 15) = .002 + .004 + .011 = .017$

b) $P(x \geq 16)$ when $n = 25$ and $\pi = .7$.
From the binomial tables $P(x \geq 16) = .134 + .165 + .171 + .148 + .103 + .057 + .024 + .007 + .002$
$= .811$

$P(x \geq 16)$ when $n = 25$ and $\pi = .6$.
From the binomial tables $P(x \geq 16) = .151 + .120 + .080 + .045 + .020 + .007 + .002$
$= .425$

c) The error probability in (a) decreases while that in (b) increases.

Supplementary Problems

5.49 a) Since N = 40000 and n = 5000, we have $\frac{n}{N} = \frac{5000}{40000} = .125$. Since more than 5% of the population is being sampled, the binomial distribution should not be used.

b) Now n = 100 and $\frac{n}{N} = \frac{100}{40000} = .0025$. Since less than 5% of the population is being sampled, the distribution of x can be approximated by a binomial distribution.

$$\mu = n\pi = 100 \frac{11000}{40000} = 100(.275) = 27.5$$

$$\sigma = \sqrt{n\pi(1 - \pi)} = \sqrt{100(.275)(.725)} = \sqrt{19.9375} = 4.4651$$

c) If the sample size is doubled, the standard deviation is increased by a factor of $\sqrt{2}$.

5.51 a) x = 0, 1, 2, 3
$P(x = 0, 1, 2, \text{ or } 3) = .10 + .15 + .20 + .25 = .70$

b) x = 0, 1, 2
$P(x = 0, 1, \text{ or } 2) = .10 + .15 + .20 = .45$

c) x = 3, 4, 5, 6
$P(x = 3, 4, 5 \text{ or } 6) = .25 + .20 + .06 + .04 = .55$

d) x = 2, 3, 4, 5
$P(x = 2, 3, 4, \text{ or } 5) = .20 + .25 + .20 + .06 = .71$

e) x = 2, 3, 4
$P(x = 2, 3, \text{ or } 4) = .20 + .25 + .20 = .65$

f) x = 0, 1, 2
$P(x = 0, 1, \text{ or } 2) = .10 + .15 + .20 = .45$

5.53 a) $p(2) = P(A_1 \text{ and } A_2) = .8(.8) = .64$

b) $p(3) = P(A_1 \text{ and } (\text{not } A_2) \text{ and } A_3] + P[(\text{not } A_1) \text{ and } A_2 \text{ and } A_3]$
$= .8(.2)(.8) + .2(.8)(.8) = 2(.128) = .256$

c) The fifth battery must be the second acceptable battery selected.
UUUAA, UUAUA, UAUUA, AUUUA
$p(5) = 4(.8)^2(.2)^3 = .02048$

d) $p(y) = (y-1)(.2)^{y-2}(.8)^2$ for $y = 2, 3, 4, \ldots$

5.55 a)

Arrival time (A,T)	w-value	Arrival times (A,T)	w-value
(1,1)	0	(4,1)	3
(1,2)	1	(4,2)	2
(1,3)	2	(4,3)	1
(1,4)	3	(4,4)	0
(1,5)	4	(4,5)	1
(1,6)	5	(4,6)	2
(2,1)	1	(5,1)	4
(2,2)	0	(5,2)	3
(2,3)	1	(5,3)	2
(2,4)	2	(5,4)	1
(2,5)	3	(5,5)	0
(2,6)	4	(5,6)	1
(3,1)	2	(6,1)	5
(3,2)	1	(6,2)	4
(3,3)	0	(6,3)	3
(3,4)	1	(6,4)	2
(3,5)	2	(6,5)	1
(3,6)	3	(6,6)	0

w-value	0	1	2	3	4	5
p(w)	$\frac{6}{36}$	$\frac{10}{36}$	$\frac{8}{36}$	$\frac{6}{36}$	$\frac{4}{36}$	$\frac{2}{36}$

b) $$\mu_w = 0\left(\frac{6}{36}\right) + 1\left(\frac{10}{36}\right) + 2\left(\frac{8}{36}\right) + 3\left(\frac{6}{36}\right) + 4\left(\frac{4}{36}\right) + 5\left(\frac{2}{36}\right) = \frac{70}{36} = 1.94 \text{ hours.}$$

5.57 a) $p(4) = P(L_1\ L_2\ L_3\ L_4) + P(B_1\ B_2\ B_3\ B_4) = (.6)^4 + (.4)^4 = .1296 + .0256 = .1552$

b) P(5) = P(Lygia wins exactly three of the first four and then game 5)
+ P(bob wins exactly three of the first four and then game 5)

$= 4(.6)^3(.4)^1(.6) + 4(.4)^3(.6)^1(.4)$

$= .20736 + .06144 = .2688$

c) x can take on values of 4, 5, 6, or 7.

p(6) = P(Lygia wins exactly three of the first five and then game 6)
+ P(Bob wins exactly three of the first five and then game 6)

$$= \left(\frac{5!}{3!2!}\right)(.6)^3(.4)^2(.6) + \left(\frac{5!}{3!2!}\right)(.4)^3(.6)^2(.4)$$

$$= 10(.6)^4(.4)^2 + 10(.4)^4(.6)^2 = .20736 + .09216 = .29952$$

p(7) = P(Lygia wins exactly three of the first six and then game 7)
+ P(Bob wins exactly three of the first six and then game 7)

$$= \left(\frac{6!}{3!3!}\right)(.6)^3(.4)^3(.6) + \left(\frac{6!}{3!3!}\right)(.4)^3(.6)^3(.4)$$

$$= 20(.6)^4(.4)^3 + 20(.4)^4(.6)^3 = .165888 + .110592 = .27648$$

x-value	4	5	6	7
p(x)	.1552	.2688	.29952	.27648

d) $\mu_x = 4(.1552) + 5(.2688) + 6(.29952) + 7(.27648) = .6208 + 1.344 + 1.79712 + 1.93536 = 5.69728$

5.59 x has a binomial distribution with n = 3 and $\pi = .4$

x	0	1	2	3
p(x)	.216	.432	.288	.064

5.61 a) For topic 1, P(at least half) = $P(x \geq 1) = p(1) + p(2) = 2(.5)(.5) + (.5)(.5) = .75$

For topic 2, P(at least half) = $P(x \geq 2) = p(2) + p(3) + p(4)$

$$= \left(\frac{4!}{2!2!}\right)(.5)^4 + \left(\frac{4!}{3!1!}\right)(.5)^4 + \left(\frac{4!}{4!0!}\right)(.5)^4$$

$$= (6 + 4 + 1)(.5)^4 = 11(.5)^4 = .6875$$

The student is more likely to have sufficient information for topic 1.

b) For topic 1, $P(x \geq 1) = p(1) + p(2)$

$$= \left(\frac{2!}{1!1!}\right)(.9)(.1) + \left(\frac{2!}{2!}\right)(.9)^2 = .18 + .81 = .99$$

For topic 2, $P(x \geq 2) = p(2) + p(3) + p(4)$

$$= \left(\frac{4!}{2!2!}\right)(.9)^2(.1)^2 + \left(\frac{4!}{3!1!}\right)(.9)^3(.1)^1 + \left(\frac{4!}{4!0!}\right)(.9)^4$$

$$= 6(.0081) + 4(.0729) + .6561 = .0486 + .2916 + .6561 = .9963$$

Under these conditions the student is more likely to have sufficient information for topic 2.

5.63 a) The two possible y-values are 1 and n+1.

b) $p(1) = (1-\pi)^n$
$p(n+1) = 1 - (1-\pi)^n$

c)

y	1	4
p(y)	$(.9)^3 = .729$	$1 - .729 = .271$

$\mu = 1(.729) + 4(.271) = .729 + 1.084 = 1.813$

Without group testing, three tests would be required. With group testing, the mean number of tests required for a group of three is 1.813.

5.65 a) x is a binomial random variable with n = 25 and $\pi = .1$. From the tables, $P(2 \leq x \leq 6) = .266 + .227 + .138 + .065 + .024 = .720$.

b) x is a binomial random variable with n = 100 and $\pi = .1$.

$\mu = 100(.1) = 10 \qquad \sigma = \sqrt{100(.1)(.9)} = 3$

c) x is a binomial random variable with n = 25 and $\pi = .1$.
$P(x \geq 7) = .007 + .002 = .009$

d) x is a binomial random variable with n = 25 and $\pi = .2$.
$P(x \leq 6) = .004 + .023 + .071 + .136 + .187 + .196 + .163 = 0.78$

5.67 a) $$p(3) = \frac{\binom{8}{3}\binom{12}{2}}{\binom{20}{5}} = \frac{\frac{8!}{3!5!}\ \frac{12!}{2!10!}}{\frac{20!}{5!15!}} = \frac{\frac{8(7)(6)}{3(2)}\ \frac{12(11)}{2}}{\frac{20(19)(18)(17)(16)}{5(4)(3)(2)}}$$

$$= \frac{56(66)}{15504} = \frac{3696}{15504} = .2384$$

b) $$p(x \le 1) = p(0) + p(1) = \frac{\binom{8}{0}\binom{12}{5}}{\binom{20}{5}} + \frac{\binom{8}{1}\binom{12}{4}}{\binom{20}{5}} = \frac{\frac{12!}{5!7!} + 8\,\frac{12!}{4!8!}}{15504}$$

$$= \frac{\frac{12(11)(10)(9)(8)}{5(4)(3)(2)} + \frac{8(12)(11)(10)(9)}{4(3)(2)}}{15504}$$

$$= \frac{792 + 3960}{15504} = \frac{4752}{15504} = .3065$$

Chapter 6
Continuous Probability Distributions

Section 1

6.1 a)

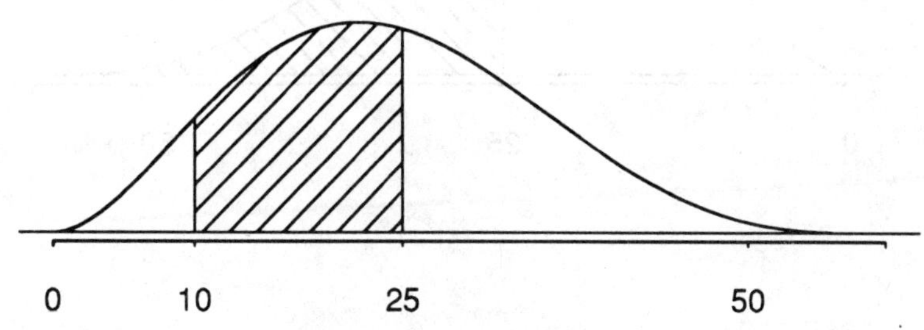

b)

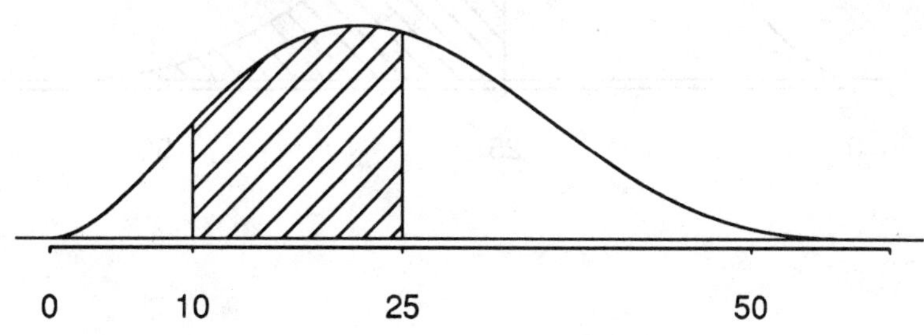

c)

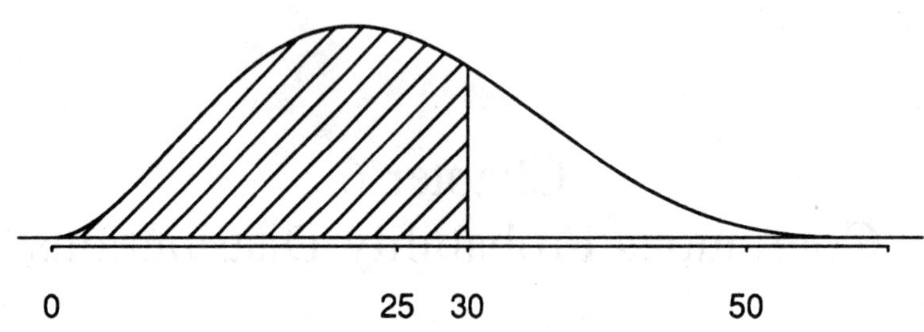

d)

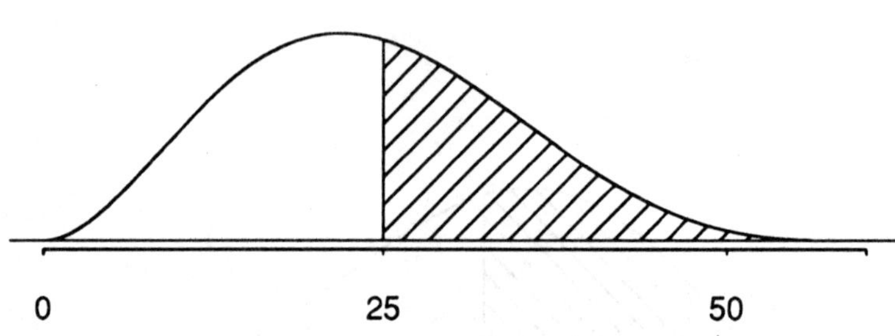

e)

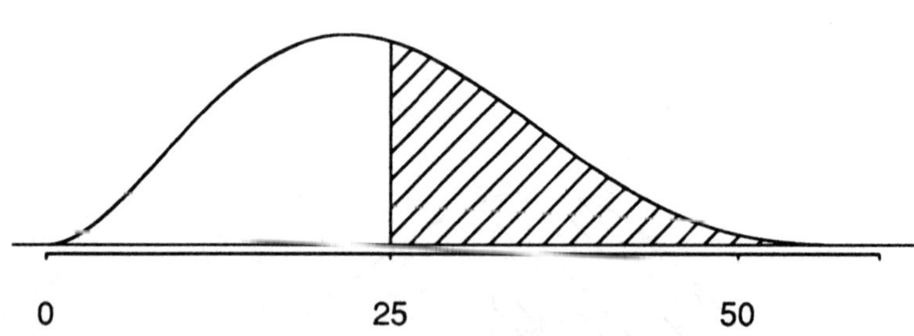

f)

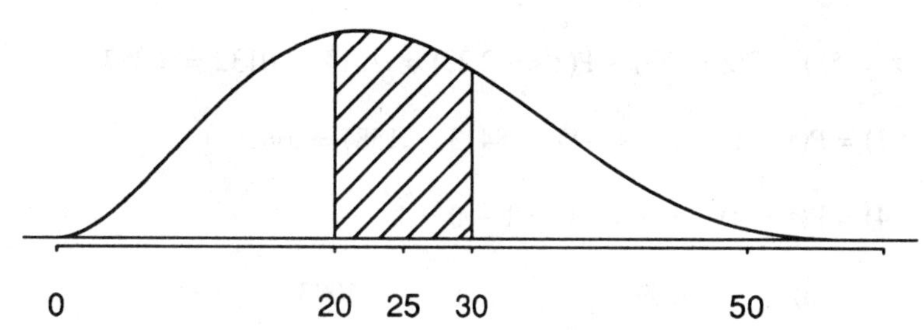

6.3 a) $\sigma = \dfrac{(10 - 0)}{\sqrt{12}} = 2.8868$

b) $P(5 - 2.8868 \le x \le 5 + 2.8868) = (1/10)(7.8868 - 2.1132) = (1/10)(5.7736) = .57736$

6.5 a) $P(x < 1/2) = 1 - P(x > 1/2) = 1 - (1/2)(1/2)(1) = .75$

b) $P(x \le 1/2) = P(x < 1/2) = .75$

c) $P(x < 1/4) = 1 - P(x > 1/4) = 1 - (1/2)(3/4)(3/2) = 1 - 9/16 = 7/16$

d) $P(1/4 < x < 1/2) = P(x > 1/4) - P(x > 1/2) = (1/2)(3/4)(3/2) - (1/2)(1/2)(1) = 9/16 - 1/4 = 5/16$

e) $P(x > 1/2) = (1/2)(1/2)(1) = 1/4$

f) $P(x \ge 3/4) = (1/2)(1/4)(1/2) = 1/16$

6.7 a) Total area = $1/2(40)(.05) = 1$

b) $P(w \le 20) = 1/2(20)(.05) = .5$
$P(w \le 10) = 1/2(10)(.05/2) = .125$
$P(w \ge 30) = 1/2(10)(.05/2) = .125$

c) $P(10 \le w \le 30) = 1 - P(x < 10 \text{ or } x > 30)$
From part (b), $P(x < 10) = .125$ and $P(x > 30) = .125$
Hence, $P(10 \le w \le 30) = 1 - (.125 + .125) = .75$

Section 2

6.9 a) $P(z < 1.75) = .9599$

b) $P(z < -.68) = .2483$

c) $P(z > 1.20) = 1 - P(z < 1.20) = 1 - .8849 = .1151$

d) $P(z > -2.82) = 1 - P(z < -2.82) = 1 - .0024 = .9976$

e) $P(-2.22 < z < .53) = P(z < .53) - P(z < -2.22) = .7019 - .0132 = .6887$

f) $P(-1 < z < 1) = P(z < 1) - P(z < -1) = .8413 - .1587 = .6826$

g) $P(-4 < z < 4) = P(z < 4) - P(z < -4) \approx 1 - 0 = 1$

6.11 a) .9909 b) .9909 c) .1093

d) $P(1.14 < z < 3.35) = P(z < 3.35) - P(z \le 1.14) = .9996 - .8729 = .1267$

e) $P(-.77 \le z \le -.55) = P(z \le -.55) - P(z < -.77) = .2912 - .2206 = .0706$

f) $P(2 < z) = 1 - P(z \le 2) = 1 - .9772 = .0228$

g) $P(-3.38 \le z) = 1 - P(z < -3.38) = 1 - .0004 = .9996$

h) $P(z < 4.98) \approx 1$

6.13 a) $c = .23$ b) $c = -.23$ c) $c = 2.75$

d) $c = 1.16$

6.15 a) $c = 1.96$ b) $c = 1.28$ c) $c = -1.28$

d) $c = 2.58$ e) $c = 2.58$ f) $c = 3.10$

6.17 a) $P(x < 5.0) = P(z < (5 - 5)/.2) = P(z < 0\ ) = .5$

b) $P(x < 5.4) = P(z < (5.4 - 5)/.2) = P(z < 2) = .9772$

c) $P(x \le 5.4) = P(z \le (5.4 - 5)/.2) = P(z \le 2) = .9772$

d) $P(4.6 < x < 5.2) = P((4.6 - 5)/.2 < z < (5.2 - 5)/.2) = P(-2 < z < 1) = P(z < 1) - P(z < -2)$
$= .8413 - .0228 = .8185$

e) $P(4.5 < x) = P((4.5 - 5)/.2 < z) = P(-2.5 < z) = 1 - P(z < -2.5) = 1 - .0062 = .9938$

f) $P(4.0 < x) = P((4 - 5)/.2 < z) = P(-5 < z) = 1 - P(z < -5) \approx 1 - 0 = 1$

6.19 a) $P(x < 14.8) = P(z < (14.8 - 15)/.1) = P(z < -2) = .0228$

b) $P(14.7 < x < 15.1) = P((14.7 - 15)/.1 < z < (15.1 - 15)/.1) = P(-3 < z < 1) = .8413 - .0013 = .8400$

6.21 a) $P(x \le 17) = P(z < (17 - 15)/1.25) = P(z < 1.6) = .9452$

b) $P(x < 17) = P(z < 1.6) = .9452$

c) $P(12 < x < 17) = P((12 - 15)/1.25 < z < (17 - 15)/1.25) = P(-2.4 < z < 1.6) = .9452 - .0082 = .9370$

d) $P(x < 15 - 2(1.25) \text{ or } x > 15 + 2(1.25)) = P(z < -2 \text{ or } z > 2) = P(z < -2) + [1 - P(z < 2)]$
$= .0228 + 1 - .9772 = .0456$

6.23 a) $c = 30 - 2.33(.6) = 30 - 1.398 = 28.602$

b) $[(30 + c) - 30]/.6 = 1.96 \Rightarrow c = 1.96(.6) = 1.176$

6.25 a) $P(x > 30.5) = P(z > (30.5 - 31)/.2) = P(z > -2.5) = 1 - P(z < -2.5) = 1 - .0062 = .9938$

b) $P(30.5 < x < 31.5) = P((30.5 - 31)/.2 < z < (31.5 - 31)/.2) = P(-2.5 < z < 2.5) = .9938 - .0062 = .9876$

$P(30 < x < 32) = P((30 - 31)/.2 < z < (32 - 31)/.2) = P(-5 < z < 5) \approx 1 - 0 = 1$

c) P(one tire being underinflated) = $P(x < 30.4) = P(z < (30.4 - 31)/.2) = P(z < -3) = .0013$

P(at least one of the four tires is underinflated) = 1 − P(none are underinflated)
$= 1 - (.9987)^4 = 1 - .9948 = .0052$

6.27 $(c - 120)/20 = -1.28 \Rightarrow c = 120 - 1.28(20) = 94.4$
Task times of 94.4 seconds or less qualify an individual for the training.

6.29 a) $P(x \le 60) = P(z \le (60 - 60)/15) = P(z \le 0) = .5$
$P(x < 60) = P(z < 0) = .5$

b) $P(45 < x < 90) = P((45 - 60)/15 < z < (90 - 60)/15) = P(-1 < z < 2) = .9772 - .1587 = .8185$

c) $P(x \ge 105) = P(z \ge (105 - 60)/15) = P(z \ge 3) = 1 - P(z < 3) = 1 - .9987 = .0013$

The probability of a typist in this population having a net rate in excess of 105 is only .0013. Hence it would be surprising if a randomly selected typist from this population had a net rate in excess of 105.

d) $P(x > 75) = P(z > (75 - 60)/15) = P(z > 1) = 1 - P(z < 1) = 1 - .8413 = .1587$
P(both exceed 75) = $(.1587)(.1587) = .0252$

Section 3

6.31 a) $P(x = 100) \approx P((99.5 - 100)/15 < z < (100.5 - 100)/15) = P(-.03 < z < .03) = .512 - .488 = .0240$

b) $P(x \le 110) \approx P(z < (110.5 - 100)/15) = P(z < .7) = .7580$

c) $P(x < 110) = P(x \le 109) = P(z < (109.5 - 100)/15) = P(z < .63) = .7357$

d) $P(75 \le x \le 125) \approx P((74.5 - 100)/15 < z < (125.5 - 100)/15) = P(-1.7 < z < 1.7)$
$= .9554 - .0446 = .9108$

6.33 a) $P(650 \le x) \approx P((649.5 - 500)/75 \le z) = P(1.99 < z) = 1 - P(z < 1.99) = 1 - .9767 = .0233$

b) $P(400 < x < 550) = P(401 \le x \le 549) \approx P((400.5 - 500)/75 < z < (549.5 - 500)/75)$
$= P(-1.33 < z < 0.66) = .7454 - .0918 = .6536$

c) $P(400 \le x \le 550) \approx P((399.5 - 500)/75 < z < (550.5 - 500)/75) = P(-1.34 < z < .67)$
$= .7486 - .0901 = .6585$

6.35 $\mu = 100(.7) = 70$ $\sigma = \sqrt{100(.7)(.3)} = \sqrt{21} = 4.5826$

a) $P(x \le 75) \approx P(z < (75.5 - 70)/4.5826 < z) = P(z < 1.2) = .8849$

b) $P(60 \le x \le 75) \approx P((59.5 - 70)/4.5826) < z < (75.5 - 70)/4.5826) = P(-2.29 < z < 1.2)$
$= .8849 - .0110 = .8739$

c) $P(80 < x) \approx 1 - P(z < 80.5 - 70)/4.5826) = 1 - P(z < 2.29) = 1 - .9890 = .0110$

d) P(at most 30 are not mountain bikes) = P(at least 70 are mountain bikes) = $P(70 \le x)$
$\approx 1 - P(z < (69.5 - 70)/4.5826) = 1 - P(z < -.11) = 1 - .4562 = .5438$

6.37 $\mu = 225(.65) = 146.25$ $\sigma = \sqrt{225(.65)(.35)} = 7.1545$

a) $P(150 \le x) \approx P((149.5 - 146.25)/7.1545 < z) = P(.45 < z) = 1 - P(z < .45) = 1 - .6736 = .3264$

b) $P(150 < x) \approx P((150.5 - 146.25)/7.1545 < z) = P(.59 < z) = 1 - P(z < .59) = 1 - .7224 = .2776$

c) $P(x < 125) \approx P(z < (124.5 - 146.25)/7.1545) = P(z < -3.04) = .0012$

6.39 $\mu = 400(.2) = 80$ $\sigma = \sqrt{400(.2)(.8)} = 8$

a) $P(75 \le x \le 100) \approx P((74.5 - 80)/8 < z < (100.5 - 80)/8) = P(-.69 < z < 2.56) = .9948 - .2451 = .7497$

b) $P(x \le 70) \approx P(z < (70.5 - 80)/8) = P(z < -1.19) = .1170$

c) $P(x < 50) \approx P(z < (49.5 - 80)/8) = P(z < -3.81) \approx 0$
If $\pi = .2$, the probability of observing fewer than 50 is essentially zero. Hence, one would question the 20% figure if fewer than 50 of 400 were actually replaced under warranty.

6.41 Due to the very distinct curvature of this plot, one would conclude that the cadmium concentration distribution is not normal. The plot suggests that the observations are coming from a skewed distribution.

6.43

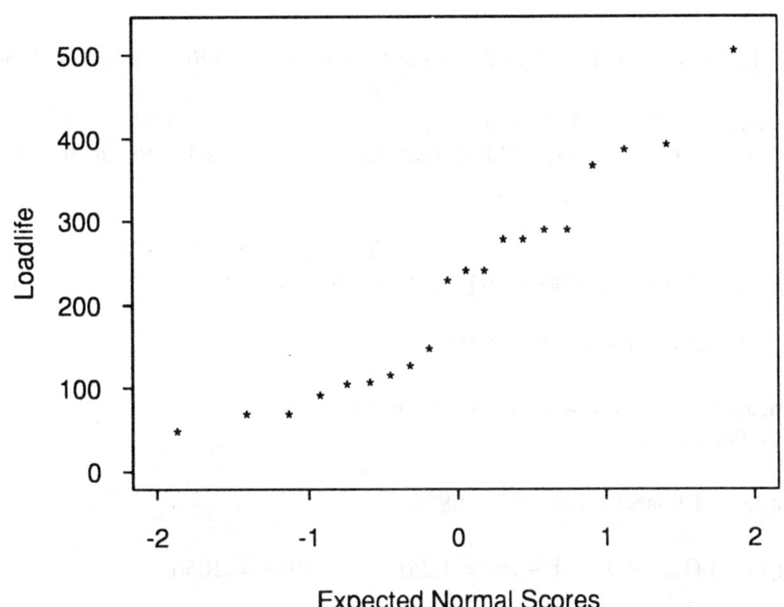

Since the graph exhibits a pattern very similar to that of a straight line, one would conclude that the distribution of the variable "load-life" could be adequately modeled by a normal distribution.

6.45

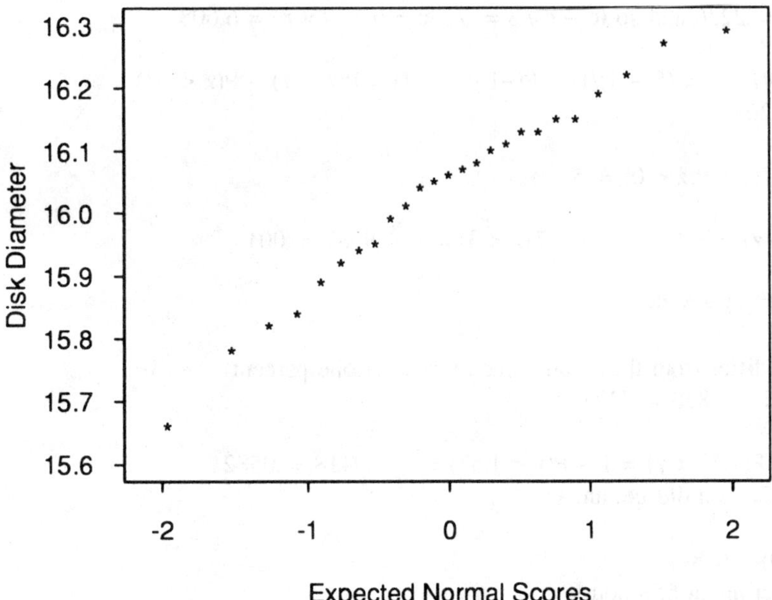

Since the graph exhibits a pattern very similar to that of a straight line, one would conclude that the distribution of the variable "disk diameter" could be adequately modeled by a normal distribution.

Supplementary Exercises

6.47 a) $P(29 < x < 31) = P((29 - 30/1.2 < z < (31 - 30)/1.2) = P(-.83 < z < .83) = .7967 - .2033 = .5934$

b) $P(x < 25) = P(z < (25 - 30)/1.2) = P(z < -4.17) \approx 0$
Yes. Based on the calculation above, it is very unlikely that, for a randomly selected car, the efficiency is less than 25 mpg.

c) $P(32 < x) = P((32 - 30)/1.2 < z) = P(1.67 < z) = 1 - P(z < 1.67) = 1 - .9525 = .0475$
P(all three have efficiencies exceeding 32 mpg) = $(.0475)^3 = .000107$

d) $c = \mu - 1.645\sigma = 30 - 1.645(1.2) = 30 - 1.974 = 28.026$

6.49 a) P(one battery lasts for at least 4 hours) = $P(4 < x) = P((4 - 6)/.8 < z) = P(-2.5 < z)$
$= 1 - P(z < -2.5) = 1 - .0062 = .9938$

P(both batteries last for at least 4 hours) = $(.9938)^2 = .9876$

b) $P(7 < x) = P((7 - 6)/.8 < z) = P(1.25 < z) = 1 - P(z < 1.25) = 1 - .8944 = .1056$

P(the cassette player works for at most 7 hours) = P(at least one battery fails before 7 hours)
= 1 − P(both last longer than 7 hours) = $1 - (.1056)^2 = 1 - .01115 = .98884$

c) We want to determine c so that P(both batteries last longer than c hours) = .05.
This requires c to be such that $P((c - 6)/.8 < z)^2 = .05$
Hence $P((c - 6)/.8 < z) = .2236$ and so $(c - 6)/.8 = .76$, $c = 6 + .76(.8) = 6.608$

6.51 a) $P(3 < x < 5) = P((3 - 4)/1 < z < (5 - 4)/1) = P(-1 < z < 1) = P(z < 1) - P(z < -1)$
$= .8413 - .1587 = .6826$

b) $P(x < 4) = P(z < (4 - 4)/1) = P(z < 0) = .5$

c) $P(7 < x) - P((7 - 4)/1 < z) = P(3 < z) = 1 - P(z < 3) = 1 - .9987 = .0013$

d) 90th percentile = 4 + 1.28(1) = 5.28

e) P(that moisture loss will differ from the mean value by at least one percent)
$= 1 - P(3 < x < 5) = 1 - .6826 = .3174$

6.53 $P(89 \leq x) = P((89 - 78)/7 < z) = P(1.57 < z) = 1 - P(z < 1.57) = 1 - .9418 = .0582$
Since you were in the upper 5.82%, you did get an A.

6.55 20th percentile = 700 + (−.84)(50) = 658
They should replace the bulbs after about 658 hours.

6.57 a) $P(250 < x < 300) = P((250 - 266)/16 < z < (300 - 266)/16) = P(-1 < z < 2.13)$
$= .9834 - .1587 = .8247$

b) $P(x < 240) = P(z < (240 - 266)/16) = P(z < -1.63) = .0516$

c) P(x is within 16 days of the mean duration) $= P(250 < x < 282) = P(-1 < z < 1)$
$= .8413 - .1587 = .6826$

d) $P(310 \le x) = P((310 - 266)/16 < z) = P(2.75 < z) = 1 - P(z < 2.75) = 1 - .9970 = .0030$
The chances of a pregnancy having a duration of at least 310 days is .003. This is a small value, so there is a bit of skepticism concerning the claim of this lady.

e) If the duration is 261 days or less, then the date of birth will be 261 + 14 = 275 days or less after coverage began. Hence the insurance company will not pay the benefits.

$P(x < 261) = P(z < (261 - 266)/16) = P(z < -.31) = .3783 \approx 38\%$

When date of conception is 14 days after coverage began, about 38% of the time the insurance company will refuse to pay the benefits because of the 275 day requirement.

6.59 For the second machine, the probability that a cork doesn't meet specifications is

$1 - P(2.9 < x < 3.1) = 1 - P((2.9 - 3.05)/.01 < z < (3.1 - 3.05)/.01) = 1 - P(-15 < z < 5) \approx 1 - 1 = 0$

Since the proportion of corks produced by this machine that will be defective is essentially zero (a much smaller value than for the first machine), I would highly recommend this second machine.

6.61 a) $P(250 \le x) \approx P((249.5 - 240)/40 < z) = P(.24 < z) = 1 - P(z < .24) = 1 - .5948 = .4052$

b) $P(250 < x) \approx P((250.5 - 240)/40 < z) = P(.26 < z) = 1 - P(z < .26) = 1 - .6026 = .3974$

c) $P(200 \le x \le 250) = P((199.5 - 240)/40 < z < (250.5 - 240)/40) = P(-1.01 < z < .26)$
$= .6026 - .1562 = .4464$

d) $P(200 \le x) \approx P((199.5 - 240)/40 < z) = P(-1.01 < z) = 1 - P(z < -1.01) = 1 - .1562 = .8438$

P(at least one has fewer than 200 pages) = 1 − P(both have 200 pages or more)
$= 1 - (.8438)^2 = 1 - .712 = .288$

6.63 $\mu = 500(.4) = 200$ $\sigma = \sqrt{500(.4)(.6)} = \sqrt{120} = 10.9545$

a) $P(250 \le x) \approx P((249.5 - 200)/10.9545 < z) = P(4.52 < z) \approx 0$

b) $P(125 \le x \le 250) \approx P((124.5 - 200)/10.9545 < z < (250.5 - 200)/10.9545)$
$= P(-6.89 < z < 4.61) \approx 1$

Chapter 7
Sampling Distributions

Section 1

7.1 A statistic is any quantity computed from the observations in a sample. A population characteristic is a quantity which describes the population from which the sample was taken.

7.3
a) population characteristic
b) statistic
c) population characteristic
d) statistic
e) statistic

7.5
a) First, obtain a list of all doctors practicing in Los Angeles County. Next, write the name of each doctor on a different card. Then shuffle the cards well, and finally, select the desired number of cards (doctors) to compose the sample.

b) First, obtain a list of all students enrolled in the university. Next, write the name of each student on a different card. Shuffle the cards well, and finally, select the desired number of cards (students) for inclusion in the sample.

c) First, obtain a listing of all possible box locations in the warehouse. Write each location on a different card. Shuffle the cards well, and then select the desired number of cards. The boxes located at those locations specified on the selected cards make up the sample.

d) Obtain a list of all registered voters in the community. Write the name of each voter on a different card. Shuffle the cards well, and select the desired number of cards (voters).

e) Obtain a list of all subscribers to the newspaper. Write the name of each subscriber on a different card. Shuffle the cards well, and select the desired number of cards (subscribers).

f) Number the radios 1 through 1000. Write the numbers 1 through 1000 on different cards, and shuffle the cards well. Select the desired number of cards (radios) for inclusion in the sample.

7.7 No unique solution.

7.9 a)

Sample	Value of t	Sample	Value of t
1, 5	6	5, 10	15
1, 10	11	5, 20	25
1, 20	21	10, 20	30

t value	6	11	15	21	25	30
probability	1/6	1/6	1/6	1/6	1/6	1/6

b) The population mean is $\mu = (1 + 5 + 10 + 20)/4 = 36/4 = 9$

$\mu_t = (1/6)(6) + (1/6)(11) + (1/6)(15) + (1/6)(21) + (1/6)(25) + (1/6)(30) = 108/6 = 18$

μ_t is twice the value of μ.

7.11

Sample	Value of Mean	Value of Median	Value of statistic #3
2, 3, 3	2.67	3	2.5
2, 3, 4	3	3	3
3, 3, 4	3.33	3	3.5
3, 4, 4	3.67	4	3.5
2, 3, 4	3	3	3
2, 3, 4	3	3	3
3, 3, 4	3.33	3	3.5
2, 3, 4	3	3	3
2, 4, 4	3.33	4	3
3, 4, 4	3.67	4	3.5

Sampling distribution of Statistic #1.

value of $\bar{x}$	2.67	3	3.33	3.67
probability	.1	.4	.3	.2

Sampling distribution of statistic #2.

value of median	3	4
probability	.7	.3

Sampling distribution of statistic #3.

value of #3	2.5	3	3.5
probability	.1	.5	.4

(Each student's discussion will differ).

7.13 a)

Sample	$\bar{x}$-value	Sample	$\bar{x}$-value
212, 379	295.5	379, 350	364.5
212, 350	281	379, 575	477
212, 575	393.5	350, 575	462.5

$\bar{x}$-value	281	295.5	364.5	393.5	462.5	477
probability	1/6	1/6	1/6	1/6	1/6	1/6

$$\mu_{\bar{x}} = (1/6)(281) + (1/6)(295.5) + (1/6)(364.5) + (1/6)(393.5) + (1/6)(462.5) + (1/6)(477)$$
$$= 46.833 + 49.25 + 60.75 + 65.583 + 77.083 + 79.5 = 379$$

b)

Sample	$\bar{x}$-value
212, 350	281
212, 575	393.5
379, 350	364.5
379, 575	477

$\bar{x}$-value	281	364.5	393.5	477
probability	.25	.25	.25	.25

$$\mu_{\bar{x}} = (1/4)(281) + (1/4)(364.5) + (1/4)(393.5) + (1/4)(477)$$
$$= 70.25 + 91.125 + 98.375 + 119.25 = 379$$

Section 2

7.15 a)

$$\sigma^2 = \frac{1}{5}[(8 - 11.8)^2+(14 - 11.8)^2+(16 - 11.8)^2+(10 - 11.8)^2+(11 - 11.8)^2]$$
$$= \frac{1}{5}[(-3.8)^2+(2.2)^2+(4.2)^2+(-1.8)^2+(-.8)^2]$$
$$= \frac{1}{5}[14.44 + 4.84 + 17.64 + 3.24 + 0.64]$$
$$= \frac{1}{5}(40.8) = 8.16$$

b) One random sample of size 2 consists of the elements 14 and 16. s^2 for this sample would be

$$(14 - 15)^2 + (16 - 15)^2 = (-1)^2 + (1)^2 = 2$$

c) Twenty-four additional samples of size 2 and their sample variances are:

Sample	Value of s^2	Sample	Value of s^2
8, 11	4.5	16, 14	2.0
14, 11	4.5	11, 8	4.5
16, 10	18.0	8, 10	2.0
10, 14	8.0	8, 16	32.0
16, 14	2.0	10, 14	8.0
10, 11	0.5	10, 14	8.0
10, 8	2.0	16, 14	2.0
11, 14	4.5	10, 8	2.0
8, 14	18.0	11, 10	0.5
10, 14	8.0	8, 11	4.5
10, 8	2.0	14, 16	2.0
16, 14	2.0	14, 16	2.0

d)

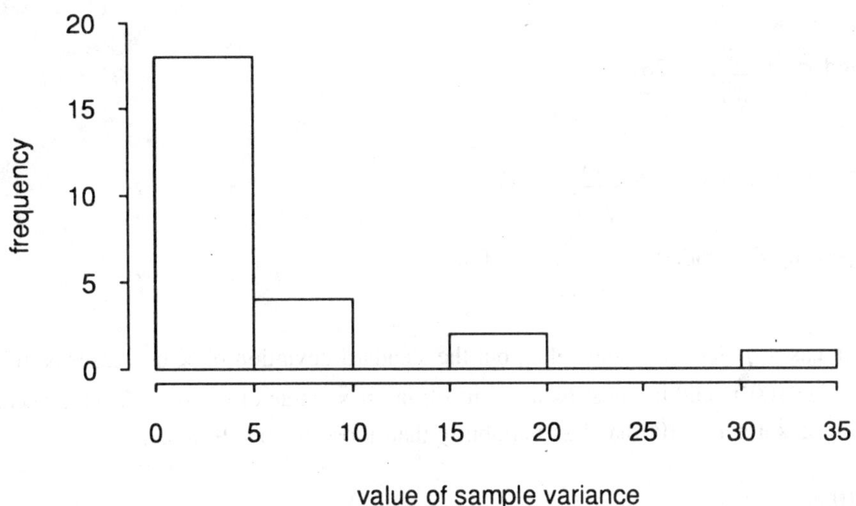

For this group of 25 samples of size two, the sample variances are either close to the population variance or smaller than the population variance. There are only three sample variances that are larger than the population variance. The sample variances do differ quite a bit in value from one sample to the next.

7.17 If samples of size ten rather than size five had been used, the histograms would be similar in that they both would be centered close to 260.25. They would differ in that the histogram based on n = 10 would have less variability than the histogram based on n = 5.

Section 3

7.19 For n = 36, 50, 100, and 400

7.21 a) $\mu_{\bar{x}} = 40,\ \sigma_{\bar{x}} = \dfrac{5}{\sqrt{64}} = \dfrac{5}{8} = .625$

Since n = 64, which exceeds 30, the shape of the sampling distribution will be approximately normal.

b) $P(\mu-.5 < \bar{x} < \mu+.5) = P(39.5 < \bar{x} < 40.5) = P((39.5 - 40)/.625 < z < (40.5 - 40)/.625)$
$= P(-.8 < z < .8) = .7881 - .2119 = .5762$

c) $P(\bar{x} < 39.3 \text{ or } \bar{x} > 40.7) = P(z < (39.3 - 40)/.625 \text{ or } z > (40.7 - 40)/.625)$
$= 1 - P(-1.12 < z < 1.12) = 1 - [.8686 - .1314] = .2628$

7.23 a) $\mu_{\bar{x}} = 2$ and $\sigma_{\bar{x}} = \dfrac{.8}{\sqrt{9}} = .267$

b) For n = 20, $\mu_{\bar{x}} = 2$ and $\sigma_{\bar{x}} = \dfrac{.8}{\sqrt{20}} = .179$

For n = 100, $\mu_{\bar{x}} = 2$ and $\sigma_{\bar{x}} = \dfrac{.8}{\sqrt{100}} = .08$

In all three cases $\mu_{\bar{x}}$ has the same value, but the standard deviation of $\bar{x}$ decreases as n increases. A sample of size 100 would be most likely to result in an $\bar{x}$ value close to μ. This is because the sampling distribution of $\bar{x}$ for n = 100 has less variability than those for n = 9 or 20.

7.25 a) $\sigma_{\bar{x}} = \dfrac{10}{\sqrt{100}} = 1$

$\bar{x}$ will be within 2 of the value of μ if $\mu-2 \le \bar{x} \le \mu+2$

$P(\mu - 2 \le \bar{x} \le \mu + 2) = P[((\mu - 2) - \mu)/1 < z < ((\mu + 2) - \mu)/1)] = P(-2 < z < 2)$
$= .9772 - .0228 = .9544$

b) Approximately 95% of the time, $\bar{x}$ will be within = 2(1) = 2 of μ.

Approximately .3% of the time, $\bar{x}$ will be further than $3\sigma_{\bar{x}} = 3(1) = 3$ from μ.

7.27 a)

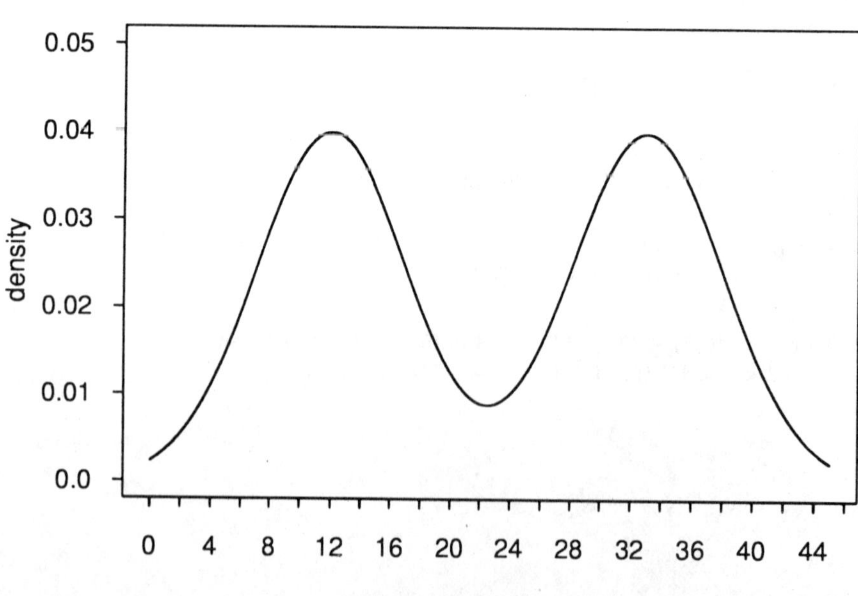

b) $\mu_{\bar{x}} = 22.0,\ \sigma_{\bar{x}} = \dfrac{16.5}{\sqrt{100}} = 1.65$

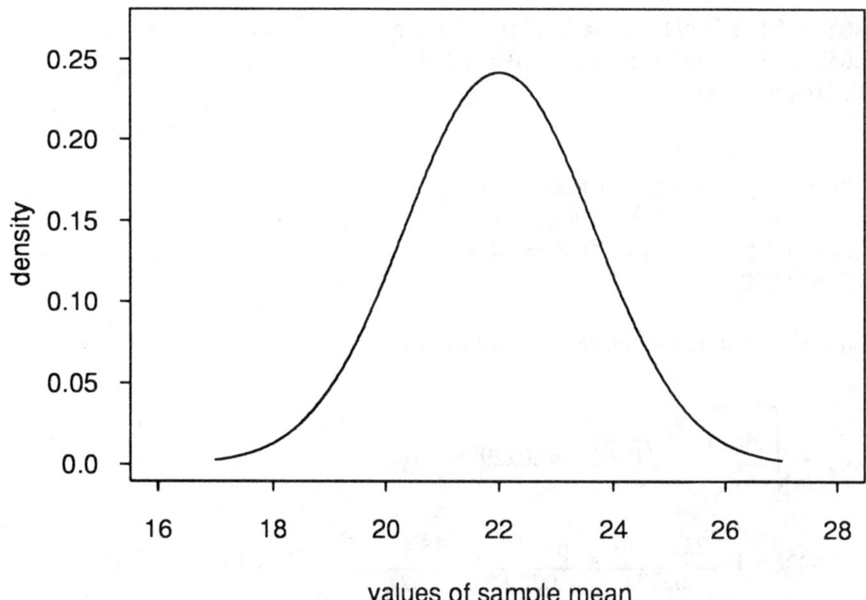

c) The central limit theorem is the justification for the following computations.

$P(\bar{x} \le 20) = P(z \le (20 - 22)/1.65) = P(z \le -1.21) = .1131$

$P(\bar{x} \ge 25) = P((25 - 22)/1.65 < z) = P(1.82 < z) = 1 - P(z \le 1.82) = 1 - .9656 = .0344$

7.29 a) Because the x distribution is normal, the $\bar{x}$ distribution will be normal (even for n less than 30).

$\mu_{\bar{x}} = 3;\ \sigma_{\bar{x}} = \dfrac{.05}{\sqrt{16}} = .0125$

b) $3 + 3\sigma_{\bar{x}} = 3 + 3(.0125) = 3.0375$ $3 - 3\sigma_{\bar{x}} = 3 - 3(.0125) = 2.9625$

c) $P(\bar{x}$ would be outside $3 \pm 3\sigma_{\bar{x}}) = 1 - (.9987 - .0013) = .0026$

d) P(the problem is detected) $= P(\bar{x} < 2.9625$ or $\bar{x} > 3.0375)$
$= P(z < (2.9625 - 3.05)/.0125$ or $z > (3.0375 - 3.05)/.0125)$
$= P(z < -7$ or $z > -1) = P(z < -7) + P(z > -1)$
$= 0 + (1 - .1587)$
$= .8413$

Section 4

7.31 When $\pi = .65$
$n = 10$, $n\pi = 10(.65) = 6.5 \geq 5$, $n(1 - \pi) = 10(.35) = 3.5 < 5$
$n = 20$, $n\pi = 20(.65) = 13 \geq 5$, $n(1 - \pi) = 20(.35) = 7 \geq 5$
So n = 20, 30, 50, 100, and 200.

When $\pi = .2$
$n = 10$, $n\pi = 10(.2) = 2 < 5$, $n(1 - \pi) = 10(.8) = 8 \geq 5$
$n = 20$, $n\pi = 20(.2) = 4 < 5$, $n(1 - \pi) = 20(.8) = 16 \geq 5$
$n = 30$, $n\pi = 30(.2) = 6 \geq 5$, $n(1 - \pi) = 30(.8) = 24 \geq 5$
So n = 30, 50, 100, and 200.

7.33 a) No, because $n\pi = 10(.3) = 3$ which does not exceed 5.

b) $\mu_p = .3 \quad \sigma_p = \sqrt{\dfrac{.3(.7)}{400}} = \sqrt{.0005} = .0229$

c) $P(.25 \leq p \leq .35) \approx P\left(\dfrac{.25 - .30}{.0229} \leq \dfrac{p - .3}{.0229} \leq \dfrac{.35 - .30}{.0229}\right) = P(-2.18 < z < 2.18)$

$= .9854 - .0146 = .9708$

d) It is smaller because as n increases, σ_p decreases. This means that the z-scores for .25 and .35 would be larger in absolute value. In fact, when $n = 500$, $\sigma_p = .0205$, and the z-scores would be $\pm$ 2.44. The area under a normal curve between -2.18 and 2.18 is smaller than the area between -2.44 and 2.44.

7.35 $\mu_p = \pi = .48, \ \sigma_p = \sqrt{\dfrac{(.48)(.52)}{500}} = .02234$

$P(.5 < p) = P((.5 - .48)/.02234 < z) = P(.90 < z) = 1 - P(z < .9) = 1 - .8159 = .1841$

Supplementary Exercises

7.37 $\mu_{\bar{x}} = .8$ and $\sigma_{\bar{x}} = \dfrac{.1}{\sqrt{100}} = .01$

$P(\bar{x} < .79) = P(z < (.79 - .8)/.01) = P(z < -1) = .1587$
$P(\bar{x} < .77) = P(z < (.77 - .8)/.01) = P(z < -3) = .0013$

7.39 $\pi = .20, \ n = 100, \ \mu_p = .2, \ \sigma_p = \sqrt{\dfrac{(.2)(.8)}{100}} = .04$

$P(.25 < p) = P((.25 - .20)/.04 < z) = P(1.25 < z) = 1 - P(z \leq 1.25) = 1 - .8944 = .1056$

7.41 a) P(850 < x < 1300) = P((850 − 1000)/150 < z < (1300 − 1000)/150) = P(−1 < z < 2)
= .9772 − .1587 = .8185

b) $\mu_{\bar{x}} = 1000,\ \sigma_{\bar{x}} = \dfrac{150}{\sqrt{10}} = 47.43$

P(950 < $\bar{x}$ < 1100) = P((950 − 1000)/47.53 < z < (1100 − 1000)/47.43)) = P(−1.05 < z < 2.11)
= .9826 − .1469 = .8357

P(850 < $\bar{x}$ < 1300) = P((850 − 1000/47.43 < z < (1300 − 1000)/47.43) = P(−3.16 < z < 6.33)
≈ 1 − .0008 = .9992

7.43 $\mu=100,\ \sigma = 30,\ \mu_{\bar{x}} = 100,\ \sigma_{\bar{x}} = \dfrac{30}{\sqrt{50}} = 4.2426$

P(5300 < total) = P(106 < $\bar{x}$) = P((106 − 100)/4.2426 < z) = P(1.41 < z) = 1 − P(z ≤ 1.41) = 1 − .9207 = .0793

Chapter 8
Estimation Using a Single Sample

Section 1

8.1 Statistic II would be preferred because it is unbiased and has smaller variance than the other two.

8.3 A point estimate of the proportion of all U.S. doctors who oppose the sale of organs is $\frac{184}{244} = 0.7541$

8.5 $p = \frac{245}{935} = .262$

8.7 a) The value of σ will be estimated by using the statistic s. For this sample,

$\Sigma x^2 = 1757.54$, $\Sigma x = 143.6$, $n = 12$

$$s^2 = \frac{\Sigma x^2 - \frac{(\Sigma x)^2}{n}}{n-1} = \frac{1757.54 - \frac{(143.6)^2}{12}}{12-1} = \frac{1757.54 - 1718.4133}{11}$$

$$= \frac{39.1267}{11} = 3.557 \text{ and } s = \sqrt{3.557} = 1.886$$

b) The population median will be estimated by the sample median. For the sample, the sample median equals (11.3 + 11.4)/2 = 11.35.

c) In this instance, a trimmed mean will be used. Trimming one observation from each end will yield an 8.3% trimmed mean. The trimmed mean equals 117.3/10 = 11.73

d) The point estimate of μ would be $\bar{x}$ = 11.967. From part (a), s = 1.886. Therefore the estimate of the 90th percentile is 11.967 + 1.28(1.886) = 14.381

8.9 a) $\bar{x}_J$ = 120.6

b) An estimate of the total amount of gas used by all these houses in January would be 10000(120.6) = 1,206,000 therms.

c) p = 8/10 = .8

d) Using the sample median, an estimate of the population median usage is (118 + 122)/2 = 120.

8.11 a) $p = \frac{\text{number in sample not defective}}{n} = \frac{68}{80} = .85$

b) P(system works) = P(both components work) = P(first component works)P(second component works) $= \pi(\pi) \approx (.85)(.85) = .7225$

Section 2

8.13 The values for parts a-d are found in the table for Standard Normal Probabilities (Appendix Table II)

a) 1.96
b) 1.645
c) 2.58
d) 1.28
e) 1.44 (approximately)

8.15 As the sample size increases, the width of the interval decreases. The interval (51.3, 52.7) has a width of 52.7 − 51.3 = 1.4 and the interval (49.4, 50.6) has a width of 50.6 − 49.4 = 1.2. Hence, the interval (49.4, 50.6) is based on the larger sample size.

8.17 The 95% confidence interval for μ_{ABC} is

a) $15.6 \pm (1.96)(\frac{5}{\sqrt{50}}) \Rightarrow 15.6 \pm 1.39 \Rightarrow (14.21, 16.99)$

b) For μ_{CBS}: $11.9 \pm 1.39 \Rightarrow (10.51, 13.29)$
For μ_{FOX}: $11.7 \pm 1.39 \Rightarrow (10.31, 13.09)$
For μ_{NBC}: $11.0 \pm 1.39 \Rightarrow (9.61, 12.39)$

c) Yes, because the plausible values for μ_{ABC} are larger than the plausible values for the other means. That is, μ_{ABC} is plausibly at least 14.21, while the other means are plausibly no greater than 13.29, 13.09, and 12.39.

8.19 $985 \pm 1.645 \dfrac{120}{\sqrt{36}} \Rightarrow 985 \pm 32.9 \Rightarrow (952.1, 1017.9)$

8.21 $n = 49$, $\bar{x} = 89$, $s = 14$

$$\bar{x} \pm (\text{z critical})\frac{s}{\sqrt{n}} = 89 \pm (1.96)\frac{14}{\sqrt{49}} = 89 \pm 3.92 = (85.08, 92.92)$$

8.23 a) $\bar{x} \pm (\text{z critical})\dfrac{s}{\sqrt{n}} = 348.95 \pm (1.645)\dfrac{176.9}{\sqrt{362}} = 348.95 \pm 15.29 = (333.66, 364.24)$

b) $\bar{x} \pm (\text{z critical})\dfrac{s}{\sqrt{n}} = 74.43 \pm (1.96)\dfrac{50.4}{\sqrt{362}} = 74.43 \pm 5.19 = (69.24, 79.62)$

c) Since the width of an interval is equal to $2(\text{z critical})\dfrac{s}{\sqrt{n}}$, and since the two intervals we are comparing have the same z critical values and the same n, the one with the larger s will be wider. Since the s for cleaning is larger than the s for food preparation, the interval for the average amount of time spent on cleaning will be wider than the interval for the average amount of time spent in food preparation.

d) The observations are being taken on a daily basis. One would suspect that cleaning is not done on a daily basis, but reserved for days off or other times when there is a large block of time available for cleaning, such as weekends. Hence, there would be many days when the amount of time spent cleaning is zero, or at least very small, and there would be days when the amount of time spent cleaning would be 8 - 12 hours. The large disparity in times results in a large s value.

8.25 $B = 0.1$, $\sigma = 0.8$

$$n = \left[\frac{1.96\sigma}{B}\right]^2 = \left[\frac{(1.96)(0.8)}{0.1}\right]^2 = (15.68)^2 = 245.86$$

Since a partial observation cannot be taken, n should be *rounded up* to n = 246.

8.27 a) $n = \left[\dfrac{(\text{z critical})\ \sigma}{B}\right]^2 = \left[\dfrac{1.96(10.2)}{1}\right]^2 = (19.992)^2 = 399.68$

Since a partial observation cannot be taken, n should be *rounded up* to 400. Notice that the sample standard deviation from the previous study is being used as an estimate of σ in determining the required sample size for the upcoming sample.

b) For average systolic blood pressure

$n = \left[\dfrac{1.96(9.8)}{1.5}\right]^2 = (12.8053)^2 = 163.98$, and hence, n should be at least 164.

For average diastolic blood pressure

$$n = \left[\frac{1.96(10.1)}{1.5}\right]^2 = (13.1973)^2 = 174.17 \text{ and therefore, n should be at least 175.}$$

Hence, to accomplish all these goals a sample size of n = 400 (the largest of the three computed n values) should be used in the forthcoming study.

8.29 For 90% confidence level:

$$n = \left[\frac{1.645\sigma}{B}\right]^2$$

For 98% confidence level:

$$n = \left[\frac{(2.33)\sigma}{B}\right]^2$$

Section 3

8.31 a) As the confidence level increases, the width of the confidence interval for π increases.

b) As the sample size increases, the width of the confidence interval for π decreases.

c) As the value of p gets farther from .5, either larger or smaller, the width of the confidence interval for π decreases.

8.33 a) $p = \frac{150}{500} = .30$

b) $.3 \pm 1.645\sqrt{\frac{.3(.7)}{500}} \Rightarrow .3 \pm 1.645(.0205) \Rightarrow .3 \pm .03 \rightarrow (.27, .33)$

8.35 The required sample size is

$$n = (.25)\left[\frac{1.96}{.02}\right]^2 = (.25)(9604) = 2401$$

The recommended sample size is large because (1) no "ball-park" value for p is known and (2) the desired accuracy is quite small.

8.37 n = 200, p = .67

$$p \pm (\text{z critical})\sqrt{\frac{p(1-p)}{n}} = .67 \pm 1.645\sqrt{\frac{.67(.33)}{200}} = .67 \pm .055 = (.615, .725)$$

8.39 $n = 1010,\ p = 17/1010 = .0168$

$$p \pm (\text{z critical})\sqrt{\frac{p(1-p)}{n}} = .0168 \pm 2.58\sqrt{\frac{.0168(.9832)}{1010}} = .0168 \pm .0104 = (.0064, .0272)$$

8.41 $n = .25\left[\frac{1.96}{B}\right]^2 = .25\left[\frac{1.96}{.05}\right]^2 = 384.16$; take n = 385.

Section 4

8.43
a) 90%
b) 95%
c) 95%
d) 99%
e) 1%
f) .5%
g) 5%

8.45 a) $n = 16,\ \bar{x} = 6.6,\ s = 3.2$

$$\bar{x} \pm (\text{t critical})\frac{s}{\sqrt{n}} = 6.6 \pm (1.75)\frac{3.2}{\sqrt{16}} = 6.6 \pm 1.4 = (5.2, 8.0)$$

b) $n = 16,\ \bar{x} = 40.3,\ s = 16$

$$\bar{x} \pm (\text{t critical})\frac{s}{\sqrt{n}} = 40.3 \pm (2.13)\frac{16}{\sqrt{16}} = 40.3 \pm 8.52 = (31.78, 48.82)$$

It is necessary to assume that the data for growth hormone is a random sample from a normal distibution.

8.47 The t-critical value for a 90% confidence interval when df = 10 − 1 = 9 is 1.83. From the given data, n = 10, $\Sigma x = 219$, and $\Sigma x^2 = 4949.92$. From the summary statistics, $\bar{x} = \frac{219}{10} = 21.9$

$$s^2 = \frac{4949.92 - \frac{(219)^2}{10}}{9} = \frac{4949.92 - 4796.1}{9} = \frac{153.82}{9} = 17.09$$

$s = \sqrt{17.09} = 4.134.$

The 90% confidence interval based on this sample data is

$$\bar{x} \pm (\text{t critical})\frac{s}{\sqrt{n}} \Rightarrow 21.9 \pm (1.83)\frac{4.134}{\sqrt{10}} \Rightarrow 21.9 \pm 2.39 \Rightarrow (19.51, 24.29).$$

8.49 a) $\bar{x} \pm (\text{t critical})\frac{s}{\sqrt{n}} = 428 \pm (2.13)\frac{35}{\sqrt{5}} = 428 \pm 33.34 = (394.66, 461.34)$

b) It would be narrower, since the standard deviation for the times for the food dehydrator is smaller than the standard deviation of times for the convection oven.

c) For convection oven:

$$2199 \pm (2.78)\frac{100}{\sqrt{5}} = 2199 \pm 124.33 = (2074.67, 2323.33)$$

For food dehydrator:

$$3920 \pm (2.78)\frac{170}{\sqrt{5}} = 3920 \pm 211.35 = (3708.65, 4131.35)$$

For microwave oven:

$$1431 \pm (2.78)\frac{30}{\sqrt{5}} = 1431 \pm 37.3 = (1393.7, 1468.3)$$

It appears that the microwave oven procedure is the most energy efficient, with the convection oven being second best, while the food dehydrator method is the least energy efficient.

8.51 a) $n = 16, \bar{x} = 10.106, s = .681$

$$\bar{x} \pm (\text{t critical})\frac{s}{\sqrt{n}} = 10.106 \pm (2.13)\frac{.681}{\sqrt{16}} = 10.106 \pm .363 = (9.743, 10.469)$$

b) We can be highly confident (that is, 95%) that the true alcohol content of bottles of cabernet produced by this manufacturer is between 9.743% and 10.469%. Since this interval does not include the asserted value of 11%, we can conclude that the manufacturer's claim is most likely incorrect.

Supplementary Exercises

8.53 a) $np = 400(.04) = 16$ and $n(1 - p) = 400(.96) = 384$. Since both np and n(1−p) are greater than or equal to 5, the sample size is large enough to justify the use of the large-sample confidence interval for a population proportion.

b) The 90% confidence interval based on this sample data is

$$p \pm (\text{z critical})\sqrt{\frac{p(1-p)}{n}} = .04 \pm 1.645\sqrt{\frac{.04(.96)}{400}} \Rightarrow .04 \pm .016 \Rightarrow (.024, .056)$$

With 90% confidence, it is estimated that the actual proportion of all ABA members who are African-American is between .024 and .056.

8.55 p = 445/602 = .739. The 95% confidence interval for the true proportion of California registered voters who prefer the cigarette tax increase is

$$p \pm (z\ \text{critical})\sqrt{\frac{p(1-p)}{n}} = .739 \pm 1.96\sqrt{\frac{.739(.261)}{602}} \Rightarrow .739 \pm .035 \Rightarrow (.704,\ .774)$$

8.57 $\bar{x} = \dfrac{220.32 + 199.18}{2} = 209.75$

The t- critical value for the 90% confidence interval is 1.75.

$(\text{t-critical})\dfrac{s}{\sqrt{n}} = 1.75\dfrac{s}{\sqrt{16}}$. Also, from the interval

$$(\text{t-critical})\frac{s}{\sqrt{16}} = 220.32 - 209.75 = 10.57$$

Hence, $1.75\dfrac{s}{\sqrt{16}} = 10.57 \Rightarrow s = \dfrac{10.57(\sqrt{16})}{1.75} \Rightarrow s = \dfrac{10.57(4)}{1.75} = 24.16$

The 95% confidence interval becomes

$$209.75 \pm 2.13\frac{24.16}{\sqrt{16}} \Rightarrow 209.75 \pm 12.8652 \Rightarrow (196.88,\ 222.62)$$

8.59 Let π denote the true proportion of all shoppers who purchase generic brands. The point estimate of π is p = 727/1442 = .5042. The 99% confidence interval for π is:

$$.5042 \pm (2.58)\sqrt{\frac{(.5042)(.4958)}{1442}} = .5042 \pm (2.58)(.0132) = .5042 \pm .0340 = (.4702,\ .5382)$$

8.61 Let μ denote the mean concentration of tetracycline for all dogs, if they were to receive tetracycline (55 mg/kg body weight) daily. If the distribution of concentration of tetracycline is normal, then the 95% confidence interval for μ is:

$$\bar{x} \pm (t\ \text{critical})\frac{s}{\sqrt{n}} = 138 \pm (2.26)\frac{65}{\sqrt{10}} = 138 \pm 46.45 = (91.55,\ 184.45)$$

8.63 a) Let μ denote the true mean flow rate using pentobarbital. The 95% confidence interval for μ is:

$$\bar{x} \pm (t\ \text{critical})\frac{s}{\sqrt{n}} = .99 \pm (1.96)\frac{.235}{\sqrt{191}} = .99 \pm .03 = (.96,\ 1.02)$$

b) Let μ denote the true mean flow rate using urethane. The 95% confidence interval for μ is:

$$\bar{x} \pm (\text{t critical})\frac{s}{\sqrt{n}} = 1.47 \pm (2.18)\frac{.314}{\sqrt{13}} = 1.47 \pm .19 = (1.28, 1.66)$$

This interval is wider than the one of part (a), because the sample size is smaller, and the sample standard deviation is larger.

c) Let μ denote the true mean flow rate using ketamine. The 95% confidence interval for μ is:

$$\bar{x} \pm (\text{t critical})\frac{s}{\sqrt{n}} = .99 \pm (2.13)\frac{.164}{\sqrt{16}} = .99 \pm .09 = (.90, 1.08)$$

Factors which contribute to the difference between the intervals of parts (a) and (c) are:

1) different sample sizes
2) using different procedures (t interval versus z interval)
3) different values for the sample standard deviations.

8.65 The 95% confidence interval for the population standard deviation of flow rate using pentobarbital is:

$$.235 \pm (1.96)\frac{.235}{\sqrt{2(191)}} = .235 \pm .024 = (.211, .259)$$

8.67 $\max(x_i) = 4.2, \quad \frac{\max(x_i)}{.5493} = \frac{4.2}{.5493} = 7.65$

Therefore, the confidence interval for θ is (4.2, 7.65).

Chapter 9
Hypothesis Testing Using a Single Sample

Section 1

9.1 $\bar{x} = 50$ is not a legitimate hypothesis, because $\bar{x}$ is a statistic, not a population characteristic. Hypotheses are always expressed in terms of a population characteristic, not in terms of a sample statistic.

9.3 If we use the hypothesis H_o: $\mu = 100$ versus H_a: $\mu > 100$, we are taking the position that the welds do not meet specifications, and hence, are not acceptable unless there is substantial evidence to show that the welds are good (i.e. $\mu > 100$). If we use the hypothesis H_o: $\mu = 100$ versus H_a: $\mu < 100$, we initially believe the welds to be acceptable, and hence, they will be declared unacceptable only if there is substantial evidence to show that the welds are faulty. It seems clear that we would choose the first set-up, which places the burden of proof on the welding contractor to show that the welds meet specifications.

9.5 Since the administration has decided to make the change if it can conclude that *more than* 60% of the faculty favor the change, the appropriate hypotheses are

$$H_o: \pi = .6 \quad \text{versus} \quad H_a: \pi > .6$$

Then, if H_o is rejected, it can be concluded that more than 60% of the faculty favor a change.

9.7 A majority is defined to be more than 50%. Therefore, the commissioner should test:

$$H_o: \pi = .5 \quad \text{versus} \quad H_a: \pi > .5$$

Section 2

9.9 a) They failed to reject the null hypothesis, because their conclusion "There is no evidence of increased risk of death due to cancer" for those living in areas with nuclear facilities is precisely what the null hypothesis states.

b) They would be making a type II error since a type I error is failing to reject the null hypothesis when the null is false.

c) Since the null hypothesis is the initially favored hypothesis and is presumed to be the case until it is determined to be false, if we fail to reject the null hypothesis, it is not proven to be true. There is just not sufficient evidence to refute its presumed truth.

9.11 A type I error involves concluding that the water being discharged from the power plant has a mean temperature in excess of 150° F when, in fact, the mean temperature is not greater than 150° F. A type II error is concluding that the mean temperature of water being discharged is 150° F or less when, in fact, the mean temperature is in excess of 150° F. I would consider a type II error to be the more serious. If we make a type II error, then damage to the river ecosystem will occur. Since it generally takes a long time to repair such damage, I would want to avoid this error. A type I error means that we require the power plant to take corrective action when it was not necessary to do so. However, since there are alternative power sources, the consequences of this error are basically financial in nature.

9.13 a) The manufacturer claims that the percentage of defective flares is 10%. Certainly one would not object if the proportion of defective flares is less than 10%. Thus, one's primary concern would be if the proportion of defective flares exceeds the value stated by the manufacturer.

b) A type I error entails concluding that the proportion of defective flares exceeds 10% when, in fact, the proportion is 10% or less. The consequence of this decision would be the filing of charges of false advertising against the manufacturer, who is not guilty of such actions. A type II error entails concluding that the proportion of defective flares is 10% when, in reality, the proportion is in excess of 10%. The consequence of this decision is to allow the manufacturer who is guilty of false advertising to continue bilking the consumer.

9.15 a) The manufacturer should test the hypotheses

$$H_o: \pi = .02 \quad \text{versus} \quad H_a: \pi < .02.$$

The null hypothesis is implicitly stating that the proportion of defective installations using robots is .02 or larger. That is, at least as large as for humans. In other words, they will not undertake to use robots unless it can be shown quite strongly that the defect rate *is less* for robots than for humans.

b) A type I error is changing to robots when in fact they are not superior to humans. A type II error is not changing to robots when in fact they are superior to humans.

c) Since a type I error means substantial loss to the company as well as to the human employees who would become unemployed a small α should be used. Therefore $\alpha = .01$ is preferred.

Section 3

9.17 The test statistic is $z = \dfrac{\bar{x} - 14}{\dfrac{s}{\sqrt{75}}}$.

Since we are performing a two-sided test, the significance level is first halved and the resulting value is used to obtain the z critical value.

a) $z > 1.96$ or $z < -1.96$

b) $z > 2.58$ or $z < -2.58$

c) $z > 1.645$ or $z < -1.645$

d) $z > 1.175$ or $z < -1.175$

9.19 a) $1 - .9850 = .0150$

b) $1 - .9980 = .0020$

c) $1 - .9990 = .0010$

9.21 1. Population characteristic of interest: μ = the average percentage of silicon dioxide.

2. H_o: $\mu = 5$

3. H_a: $\mu \neq 5$

4. Test statistic: $z = \dfrac{\bar{x} - 5}{\dfrac{s}{\sqrt{n}}}$

5. Rejection region: $z < -2.58$ or $z > 2.58$

6. Computations: $n = 36$, $\bar{x} = 5.21$, $s = .38$,

$$z = \frac{5.21 - 5}{\frac{.38}{\sqrt{36}}} = 3.316$$

7. Conclusion: Since 3.316 is greater than 2.58, the calculated z falls into the rejection region and H_o is rejected, using an $\alpha = .01$. There is sufficient evidence to conclude that the true mean percentage of silicon dioxide in this type of cement has a value which differs from 5.

9.23 a) 1. Population characteristic of interest: μ = average parts per million of PCB in the well's water.

2. H_0: $\mu = 5$

3. H_a: $\mu < 5$

4. Test statistic: $z = \dfrac{\bar{x} - 5}{\dfrac{s}{\sqrt{n}}}$

5. Rejection region: $z < -2.33$

6. Computations: $n = 36$, $\bar{x} = 4.82$, $s = .6$,

$$z = \frac{4.82 - 5}{\dfrac{.6}{\sqrt{36}}} = -1.80$$

7. Conclusion: Since -1.80 is greater then -2.33, the calculated z does not fall in the rejection region and H_0 is not rejected at the .01 level. There is not sufficient evidence to substantiate the claim that the water is safe.

b) A type I error is concluding that the water is safe when in fact it has too much PCB in it. Since this error would lead to severe health problems, the risk of making this error should not be taken larger than the stated level of significance .01.

9.25 1. Let μ denote the true average compressive strength for this type of brick.

2. H_0: $\mu = 3200$

3. H_a: $\mu < 3200$

4. Test statistic: $z = \dfrac{\bar{x} - 3200}{\dfrac{s}{\sqrt{n}}}$

5. Rejection region: $z < -1.645$

6. $$z = \frac{3109 - 3200}{\dfrac{156}{\sqrt{36}}} = \frac{-91}{26} = -3.50$$

7. Conclusion: Since -3.50 is less than -1.645, the calculated z does fall in the rejection region and H_0 is rejected at the .05 level. The data supports the conclusion that the mean compressive strength of this type of brick is below 3200.

9.27 a)

1. Let μ denote the true mean anger expression for frequent marijuana users.

2. H_o: $\mu = 41.5$

3. H_a: $\mu > 41.5$

4. Test statistic: $z = \dfrac{\bar{x} - 41.5}{\frac{s}{\sqrt{n}}}$

5. Rejection region: $z > 1.645$

6. Computation: $n = 47$, $\bar{x} = 42.72$, $s = 6.05$,

$$z = \frac{42.72 - 41.5}{\frac{6.05}{\sqrt{47}}} = 1.382$$

7. Conclusion: Since 1.382 does not exceed 1.645, the calculated z does not fall in the rejection region. At level of significance .05, H_o is not rejected. The data does not provide sufficient evidence to conclude that the true mean anger expression score for frequent marijuana users exceeds that for nonusers.

b) A type I error would be concluding that the mean anger expression score for frequent marijuana users exceeds that for nonusers, when in reality it does not exceed the mean anger expression score for nonusers.

A type II error would be concluding that the mean anger expression score for frequent marijuana users does not exceed that for nonusers, when in reality it does exceed the mean anger expression score for nonusers.

9.29 a)

1. Let μ denote the mean number of corrections and omissions for seven-year-old children who are reading at below grade level.

2. H_o: $\mu = 19.9$

3. H_a: $\mu \neq 19.9$

4. Test statistic: $z = \dfrac{\bar{x} - 19.9}{\frac{s}{\sqrt{n}}}$

5. Rejection region: $z < -2.58$ or $z > 2.58$

6. Computations: $n = 48$, $\bar{x} = 17.68$, $s = 4.2$,

$$z = \frac{17.68 - 19.9}{\frac{4.2}{\sqrt{48}}} = -3.66$$

7. Conclusion: Since −3.66 is less than −2.58, the computed z falls in the rejection region. At level of significance .01, H_o is rejected. The data provides sufficient evidence to conclude that the mean number of corrections and omissions for seven-year-old children who are reading below grade level differs from that of seven-year-old children who are reading at grade level.

b) If a level of significance of .05 had been used the rejection region would be: $z < -1.96$ or $z > 1.96$. Since the computed z would still fall into the rejection region, the conclusion would not have changed.

9.31 a) 1. Let μ denote the true mean number of insects per plant when Permethrin is used.

2. H_o: $\mu = 8$

3. H_a: $\mu < 8$

4. Test statistic: $z = \frac{\bar{x} - 8}{\frac{s}{\sqrt{n}}}$

5. Rejection region: $z < -1.645$

6. Computations: $n = 96$, $\bar{x} = 7.2$, $s = 6.3$,

$$z = \frac{7.2 - 8}{\frac{6.3}{\sqrt{96}}} = -1.24$$

7. Conclusion: Since −1.24 is not less than −1.645, the computed z does not fall in the rejection region. At level of significance .05, H_o is not rejected. The sample does not contain sufficient evidence to conclude that the mean number of insects per plant is less than 8 when Permethrin is used.

b) If the significance level had been .01, the rejection region would have been $z < -2.33$. The computed z would not have been in the rejection region and the conclusion would not have changed.

9.33 1. Let μ denote the mean playing time of this manufacturer's audio tapes.

2. H_o: $\mu = 90$

3. H_a: $\mu < 90$

4. Test statistic: $z = \dfrac{\bar{x} - 90}{\dfrac{s}{\sqrt{n}}}$

5. For $\alpha = 0.05$, reject H_0 if $z < -1.645$.

6. For $n = 900$, $\bar{x} = 89.95$, $s = .3$,

$$z = \frac{89.95 - 90}{\dfrac{.3}{\sqrt{900}}} = -5$$

7. Since the computed z of −5 is less than the z critical of −1.645, the computed z falls in the rejection region. H_0 is rejected, and it is concluded that the mean playing time is less than 90 minutes. Since H_0 is rejected, the results are said to be statistically significant. However, 89.95 is only .05 minutes below the claimed value of 90. The fact that the mean playing time is less than the claimed value by so little leads one to conclude that the results may not be of much practical significance.

Section 4

9.35 H_0 is rejected if P-value $\leq \alpha$. H_0 should be rejected for only the following pair:

d) P-value = .084, $\alpha = 0.10$

9.37 Since this is a two-tailed test, the P-value is equal to twice the area captured in the tail in which z falls. Using Appendix Table II, the P-values are:

a) 2(.0179) = .0358

b) 2(.0401) = .0802

c) 2(.2810) = .5620

d) 2(.0749) = .1498

e) 2(0) = 0

9.39 a) z = 3.00

b) P-value = .0014

c) H_0 is rejected if the P-value is less than α. Since the P-value of .0014 is less than α, which is 0.05, H_0 should be rejected in favor of the conclusion that the machine is overfilling.

9.41 Population characteristic of interest: μ = average drying time when additive is used in the paint

H_o: $\mu = 75$ H_a: $\mu < 75$

Test statistic: $z = \dfrac{\bar{x} - 75}{\frac{s}{\sqrt{n}}}$

Computations: $n = 100$, $\bar{x} = 68.5$, $s = 9.4$,

$$z = \frac{68.5 - 75}{\frac{9.4}{\sqrt{100}}} = -6.91$$

P-value = $P(z < -6.91) < .0002$

Conclusion: Since the P-value of .0002 is less than the chosen α of .01, H_o is rejected and the experimental evidence indicates strongly that the additive shortens the drying time.

9.43 a) Population characteristic of interest: μ = mean score on this test when students are given an "organizer".

H_o: $\mu = 32$ H_a: $\mu > 32$

Test statistic: $z = \dfrac{\bar{x} - 32}{\frac{s}{\sqrt{n}}}$

For $\alpha = .05$, reject H_o if $z > 1.645$

Computations: $n = 100$, $\bar{x} = 32.96$, $s = 9.24$,

$$z = \frac{32.96 - 32}{\frac{9.24}{\sqrt{100}}} = 1.04$$

Conclusion: Since the calculated z of 1.04 does not exceed the z critical value of 1.645, H_o is not rejected using a level of significance of .05. The data does not support the conclusion that preliminary exposure to an organizer increases the true average score on this test.

b) P-value = $P(1.04 < z) = 1 - P(z \le 1.04) = 1 - .8505 = .1492$
The null hypothesis would be rejected for significance levels that are .1492 or larger.

Section 5

9.45 1. Let π denote the proportion of all pizzas ordered that are large.

2. H_o: $\pi = .75$

3. H_a: $\pi > .75$

4. Test statistic: $z = \dfrac{p - .75}{\sqrt{\dfrac{.75(.25)}{n}}}$

5. Rejection region: $z > 1.28$

6. Computations: $n = 150$, $p = \dfrac{120}{150} = .80$,

$$z = \frac{.80 - .75}{\sqrt{\dfrac{.75(.25)}{150}}} = 1.41$$

7. Since 1.41 exceeds 1.28, the computed z falls in the rejection region. At level of significance .10, H_o is rejected. The sample provides sufficient evidence to conclude that the true proportion of pizzas ordered that are of the large size exceeds .75.

9.47 1. Let π represent the proportion of local residents who oppose hunting on Morro Bay

2. H_o: $\pi = .50$

3. H_a: $\pi > .50$

4. Since $n\pi = 750(.50) = 375 \geq 5$, and $n(1 - \pi) = 750(.5) = 375 \geq 5$, the large sample z test may be used.

$$z = \frac{p - .5}{\sqrt{\dfrac{.5(.5)}{n}}}$$

5. For $\alpha = .01$, reject H_o if $z > 2.33$.

6. $p = \dfrac{560}{750} = .7467$

$$z = \frac{.7467 - .5}{\sqrt{\dfrac{.5(.5)}{750}}} = \frac{.2467}{.0183} = 13.51$$

7. Since the calculated z of 13.51 exceeds the z critical value of 2.33, H_o is rejected. At level of significance .01, the data supports the conclusion that the majority of local residents oppose hunting on Morro Bay.

9.49 1. Let π represent the true incidence rate of this type of chromosome defect in the U.S. adult male population.

2. H_o: $\pi = \frac{1}{80} = .0125$

3. H_a: $\pi \neq .0125$

4. Since $n\pi = 600(\frac{1}{80}) = 7.5 \geq 5$, and $n(1 - \pi) = 600(\frac{79}{80}) = 592.5 \geq 5$, the large sample z test for π may be used.

$$z = \frac{p - .0125}{\sqrt{\frac{.0125(.9875)}{600}}} = \frac{p - .0125}{.004536}$$

5. For $\alpha = 0.05$, reject H_o if $z > 1.96$ or $z < -1.96$.

6. $p = \frac{12}{600} = .0.02$

$$z = \frac{.02 - .0125}{.004536} = 1.65$$

7. Since the calculated z of 1.65 does not exceed the z critical value of 1.96, H_o is not rejected. Therefore, it cannot be concluded that the incidence rate of the defect among prisoners differs from the presumed rate for the entire adult population.

Since H_o is not rejected, it is possible that a type II error has been made.

9.51 1. Let π denote the proportion of physicians who know the generic name for methadone.

2. H_o: $\pi = .5$

3. H_a: $\pi < .5$

4. Test statistic: $z = \frac{p - .5}{\sqrt{\frac{.5(.5)}{n}}}$

5. Rejection region: $z < -2.33$

6. Computations: $n = 102$, $p = \frac{47}{102} = .4608$,

$$z = \frac{.4608 - .5}{\sqrt{\frac{.5(.5)}{102}}} = -.792$$

7. Since $-.792$ is not less than -2.33, the computed z does not fall into the rejection region. At level of significance .01, H_o is not rejected. The evidence in the sample does not support the conclusion that fewer than 50% of all physicians know the generic name for methadone.

9.53 1. Let π denote the true proportion of Californians who feel that their life is stressful.

2. H_o: $\pi = .5$

3. H_a: $\pi < .5$

4. Test statistic: $z = \frac{p - .5}{\sqrt{\frac{.5(.5)}{n}}}$

5. Rejection region: $z < -1.645$

6. Computations: $n = 1008$, $p = \frac{484}{1008} = .4802$

$$z = \frac{.4802 - .5}{\sqrt{\frac{.5(.5)}{1008}}} = -1.26$$

7. Since -1.26 is not less than -1.645, the computed z does not fall into the rejection region. At level of significance .05, H_o is not rejected. The sample does not contain sufficient evidence to conclude that fewer than half of Californians feel that their life is stressful.

9.55 a) 1. Let π denote the proportion of all doctors who feel it is sometimes appropriate to help a seriously ill person die.

2. H_o: $\pi = .6$

3. H_a: $\pi > .6$

4. Test statistic: $z = \frac{p - .6}{\sqrt{\frac{.6(.4)}{n}}}$

5. Rejection region: $z > 2.33$

6. Computations: $n = 588$, $p = \frac{365}{588} = .6207$

$$z = \frac{.6207 - .6}{\sqrt{\frac{.6(.4)}{588}}} = 1.03$$

7. Since 1.03 does not exceed 2.33, the computed z does not fall into the rejection region. At level of significance, H_0 is not rejected. Based on this sample, it cannot be concluded that more than 60% of all doctors feel it is sometimes appropriate to help a seriously ill person die.

b) At level of significance .05, the rejection region would be: $z > 1.645$. Since 1.03 does not exceed 1.645, the conclusion would not change.

Section 6

9.57

	Test Statistic	d.f.	Rejection Region
a)	$t = \frac{\bar{x} - 30}{\frac{s}{\sqrt{10}}}$	9	Reject H_0 if $t > 2.26$ or $t < -2.26$
b)	$t = \frac{\bar{x} - 30}{\frac{s}{\sqrt{18}}}$	17	Reject H_0 if $t > 2.90$ or $t < -2.90$
c)	$t = \frac{\bar{x} - 30}{\frac{s}{\sqrt{25}}}$	24	Reject H_0 if $t > 3.75$ or $t < -3.75$
d)	$t = \frac{\bar{x} - 30}{\frac{s}{\sqrt{50}}}$	49	Reject H_0 if $t > 1.67$ or $t < -1.67$

9.59 Population characteristic of interest: μ= mean maximum weight of lift (in kg).

H_0: $\mu = 25$ H_a : $\mu > 25$

Test statistic: $t = \frac{\bar{x} - 25}{\frac{s}{\sqrt{n}}}$ with d.f. = 4

For $\alpha = 0.05$, reject H_o if $t > 2.13$

$$t = \frac{27.54 - 25}{5.47/\sqrt{5}} = 1.038$$

The calculated t of 1.038 does not exceed the critical t of 2.13 and so there is insufficient evidence to reject H_o. It cannot be concluded that the mean maximum weight of lift exceeds 25 kilograms.

9.61 Let μ = mean heat flux level when coal dust is used.

H_o: $\mu = 29.0$ H_a: $\mu > 29.0$

Test statistic: $t = \dfrac{\bar{x} - 29.0}{\frac{s}{\sqrt{n}}}$ with d.f. = 8 − 1 = 7

Rejection region: Reject H_o if $t > 1.90$.

From the Minitab output, the calculated t = .77. This value does not fall into the rejection region, so at level of significance .05, H_o is not rejected. The sample data suggests that the mean heat flux level when coal dust is used is not greater than when grass is used.

9.63 Let μ denote the mean pH value of the soil when treated with a solution of 25% water and 75% effluent.

H_o: $\mu = 8.75$ H_a: $\mu < 8.75$

$t = \dfrac{\bar{x} - 8.75}{\frac{s}{\sqrt{n}}}$ with d.f. = 4

For $\alpha = .01$, reject H_o if $t < -3.75$.

$$t = \frac{8 - 8.75}{\frac{.05}{\sqrt{5}}} = -33.54$$

Since the calculated t of −33.54 falls in the rejection region, H_o is rejected. There is sufficient evidence to conclude that the mean pH is lower than 8.75.

9.65 First, note that this is a one-tail test, so P-value = P(t > observed t).

a) .05 > P-value > .025

b) .0005 > P-value

c) .025 > P-value > .01

d) P-value > .10

e) .01 > P-value > .001

9.67 a) μ = mean tensile strength for springs using the roller method.

H_o: $\mu = 2150$ versus H_a: $\mu > 2150$

b) $t = \dfrac{\bar{x} - 2150}{\frac{s}{\sqrt{n}}}$ with d.f. = 16 − 1 = 15

c) $t = \dfrac{2160 - 2150}{\frac{30}{\sqrt{16}}} = 1.33$

d) P-value > .10

e) H_o would not be rejected, since the P-value exceeds .05.

Supplementary Exercises

9.69 Let μ denote the mean systolic blood pressure for all clerical workers at this business.

H_o: $\mu = 110$ H_a: $\mu > 110$

$$z = \frac{\bar{x} - 110}{\frac{s}{\sqrt{n}}}$$

At level .01, reject H_o if $z > 2.33$.

$$z = \frac{111.63 - 110}{\frac{11.94}{\sqrt{65}}} = 1.10$$

The computed z does not fall in the rejection region, so H_o is not rejected. There is not sufficient evidence to conclude that the mean systolic blood pressure for all clerical workers at this business exceeds 110.

9.71 a) Daily caffeine consumption cannot be a negative value. Since the standard deviation is larger than the mean, this would imply that a sizable portion of a normal curve with this mean and this standard deviation would extend into the negative values on the number line. Therefore, it is not plausible that the population distribution of daily caffeine consumption is normal.

Since the sample size is large (greater than 30) the Central Limit Theorem allows for the conclusion that the distribution of $\overline{x}$ is approximately normal even though the population distribution is not normal. So it is not necessary to assume that the population distribution of daily caffeine consumption is normal to test hypotheses about the value of population mean consumption.

b) Let μ denote the population mean daily consumption of caffeine.

$H_o : \mu = 200 \qquad H_a : \mu > 200$

$$z = \frac{\overline{x} - 200}{s/\sqrt{n}}$$

Reject H_o if P-value < .10

$$z = \frac{\overline{x} - 200}{s/\sqrt{n}} = \frac{215 - 200}{235/\sqrt{47}} = 0.44$$

P-value = P(z > 0.44) = 1 − .6700 = .33

Since the P-value exceeds the level of significance of .10, H_o is not rejected. The data does not support the conclusion that the population mean daily caffeine consumption exceeds 200 mg.

9.73 a) Let π represent the response rate when the distributor is stigmatized by an eye patch.

$H_o: \pi = .40 \qquad H_a: \pi > .40$

The test statistic is: $z = \dfrac{p - .40}{\sqrt{\dfrac{(.4)(.60)}{n}}}$.

For $\alpha = 0.05$, reject H_o if z > 1.645.

From the data: n = 200, $p = \dfrac{109}{200} = .545$,

$$z = \frac{.545 - .40}{\sqrt{\dfrac{(.40)(.60)}{200}}} = 4.19$$

Since the calculated z of 4.19 exceeds the z critical value of 1.645, the null hypothesis is rejected. The data strongly suggests that the response rate does exceed the rate in the past.

b) P-value = P(z > 4.19) < .0002. Since the P-value of .0002 is less than the α value of .05, H_o is rejected.

9.75 Let π denote the proportion of senior citizens who are satisfied with their grocery purchases.

H_o: $\pi = .8$ H_a: $\pi < .8$

The test statistic is: $z = \dfrac{p - .8}{\sqrt{\dfrac{(.8)(.2)}{n}}}$.

For $\alpha = .01$, reject H_o if $z < -2.33$.

From the sample: $p = \dfrac{270}{404} = .6683$,

$$z = \frac{.6683 - .8}{\sqrt{\dfrac{(.8)(.2)}{404}}} = -6.62$$

Since the calculated z of −6.62 is less than the z critical of −2.33, the null hypothesis is rejected. The data strongly suggests that the proportion of senior citizens who are satisfied with their grocery purchases is smaller than that for the general population.

9.77 Let μ denote the mean met-enkephalin level of SIDS victims.

H_o: $\mu = 7.48$ H_a: $\mu > 7.48$

The test statistic is: $t = \dfrac{\bar{x} - 7.48}{\dfrac{s}{\sqrt{n}}}$ with d.f. = 11.

For $\alpha = .05$, reject H_o if $t > 1.80$.

Using the information from the sample,

$$t = \frac{7.66 - 7.48}{\dfrac{3.78}{\sqrt{12}}} = 0.16.$$

Since the t calculated value of 0.16 does not fall in the rejection region, the null hypothesis is not rejected. The data does not allow one to conclude that mean met-enkephalin level of SIDS victims is higher than that of children not suffering from SIDS.

9.79 Let μ denote the mean root length of pearl millet when irrigated with the 50% wastewater solution.

H_o: $\mu = 6.40$ H_a: $\mu \neq 6.40$

The test statistic is: $z = \dfrac{\bar{x} - 6.40}{\frac{s}{\sqrt{n}}}$.

For $\alpha = .05$, reject H_o if $z < -1.96$ or $z > 1.96$.

Using the sample data: $z = \dfrac{4.76 - 6.4}{\frac{.48}{\sqrt{40}}} = -21.6$.

The calculated z value of −21.6 falls in the rejection region, hence, the null hypothesis is rejected. The data strongly suggests that irrigation with the wastewater solution results in a mean root length that differs from 6.40.

9.81 a) Let μ denote the true mean time required to achieve 100° F with the heating equipment of this manufacturer.

H_o: $\mu = 15$ H_a: $\mu > 15$

The test statistic is: $z = \dfrac{\bar{x} - 15}{\frac{s}{\sqrt{n}}}$.

For $\alpha = 0.05$, reject H_o if $z > 1.645$.

From the sample: $n = 32$, $\bar{x} = 17.5$, $s = 2.2$,

$$z = \frac{17.5 - 15}{\frac{2.2}{\sqrt{32}}} = 6.43.$$

The calculated z value of 6.43 falls in the rejection region, the null hypothesis is rejected. The data does cast doubt on the company's claim that it requires at most 15 minutes to achieve 100° F.

b) P-value = $P(z > 6.43) < .0002$. Because the P-value is smaller than α, H_o is rejected.

Chapter 10
Comparing Two Populations or Treatments

Section 1

10.1 $\mu_{\bar{x}_1 - \bar{x}_2} = \mu_1 - \mu_2 = 30 - 25 = 5$

$$\sigma_{\bar{x}_1 - \bar{x}_2} = \sqrt{\frac{\sigma_1^2}{n_1} + \frac{\sigma_2^2}{n_2}} = \sqrt{\frac{(2)^2}{40} + \frac{(3)^2}{50}} = \sqrt{\frac{4}{40} + \frac{9}{50}} = \sqrt{.28} = .529$$

Since both n_1 and n_2 are large, the sampling distribution of $\bar{x}_1 - \bar{x}_2$ is approximately normal. It is centered at 5 and the standard deviation is .529.

10.3 a) H_o: $\mu_1 - \mu_2 = 0$ $\quad H_a$: $\mu_1 - \mu_2 \neq 0$

b) H_o: $\mu_1 - \mu_2 = 0$ $\quad H_a$: $\mu_1 - \mu_2 < 0$

10.5 H_o: $\mu_1 - \mu_2 = 0$ $\quad H_a$: $\mu_1 - \mu_2 \neq 0$

$$z = \frac{(\bar{x}_1 - \bar{x}_2) - 0}{\sqrt{\frac{s_1^2}{n_1} + \frac{s_2^2}{n_2}}}$$

For $\alpha = 0.05$, reject H_o if $z > 1.96$ or $z < -1.96$.

$$z = \frac{(36500 - 33400) - 0}{\sqrt{\frac{(2200)^2}{40} + \frac{(1900)^2}{40}}} = \frac{3100}{\sqrt{121000 + 90250}} = \frac{3100}{\sqrt{211250}} = \frac{3100}{459.62} = 6.74$$

Since the calculated z of 6.74 exceeds the z critical value 1.96, H_o is rejected. It is concluded that the two brands of FR78-15 tires do not have equal mean tread lives. The data strongly suggests that brand 1 has a larger mean tread life than brand 2.

10.7 Let μ_1 and μ_2 denote the true average counts for two-year and four-year old horses, respectively.

H_o: $\mu_1 - \mu_2 = 0$ H_a: $\mu_1 - \mu_2 < 0$

$$z = \frac{(\bar{x}_1 - \bar{x}_2) - 0}{\sqrt{\frac{s_1^2}{n_1} + \frac{s_2^2}{n_2}}}$$

For $\alpha = 0.001$, reject H_o if $z < -3.09$.

$$z = \frac{(51 - 56) - 0}{\sqrt{\frac{(5.6)^2}{197} + \frac{(4.3)^2}{77}}} = \frac{-5}{\sqrt{.4}} = \frac{-5}{.63} = -7.91$$

Since the calculated z of −7.91 is less than the z critical value −3.09, H_o is rejected. The data strongly suggests that the average neutrophil count for four-year olds exceeds that for two-year olds.

10.9 Let μ_1 denote the true mean shoulder extension for 45-75 year-old males and μ_2 the true mean shoulder extension for 45-75 year-old females.

H_o: $\mu_1 - \mu_2 = 0$ H_a: $\mu_1 - \mu_2 \neq 0$

$$z = \frac{(\bar{x}_1 - \bar{x}_2) - 0}{\sqrt{\frac{s_1^2}{n_1} + \frac{s_2^2}{n_2}}}$$

For $\alpha = 0.01$, reject H_o if $z > 2.58$ or $z < -2.58$.

$$z = \frac{(12.0 - 10.3) - 0}{\sqrt{\frac{(4.9)^2}{39} + \frac{(3.7)^2}{41}}} = \frac{1.70}{0.9744} = 1.74$$

Since the computed z of 1.74 does not fall into the rejection region, the null hypothesis is not rejected. At level of significance 0.01, the data does not support the conclusion that the mean shoulder extension for elderly men differs from the mean shoulder extension for elderly women.

10.11 1. μ_1 = mean hostility score for children of divorced parents
μ_2 = mean hostility score for children of married parents
$\mu_1 - \mu_2$ = difference in mean hostility scores

2. H_o: $\mu_1 - \mu_2 = 0$

3. H_a: $\mu_1 - \mu_2 > 0$

4. Test statistic:

$$z = \frac{(\bar{x}_1 - \bar{x}_2) - 0}{\sqrt{\frac{s_1^2}{n_1} + \frac{s_2^2}{n_2}}}$$

5. Computations:

$$z = \frac{5.38 - 1.94 - 0}{\sqrt{\frac{(3.96)^2}{54} + \frac{(3.10)^2}{54}}} = \frac{3.44}{.684} = 5.03$$

6. Since this is an upper-tailed test, the associated P-value is the area under the z curve and to the right of the computed z value of 5.03. Since this area is approximately 0, P-value ≈ 0

7. The P-value is smaller than my selected significance level of .01, so we reject H_o. This data does support the researcher's hypothesis that the mean hostility score is higher for children of divorced parents than for children of married parents.

10.13 Let μ_1 denote the true mean final exam scores for students who take quizzes at the beginning of the class and μ_2 the true mean final exam scores for students who take quizzes at the end of class.

H_o: $\mu_1 - \mu_2 = 0$ H_a: $\mu_1 - \mu_2 \neq 0$

$$z = \frac{(\bar{x}_1 - \bar{x}_2) - 0}{\sqrt{\frac{s_1^2}{n_1} + \frac{s_2^2}{n_2}}}$$

No level of significance was specified, so $\alpha = .01$ will be used.

For $\alpha = 0.01$, reject H_o if $z > 2.58$ or $z < -2.58$.

$$z = \frac{(143.7 - 131.7) - 0}{\sqrt{\frac{(21.2)^2}{40} + \frac{(20.9)^2}{40}}} = \frac{12}{\sqrt{11.236 + 10.92}} = \frac{12}{4.71} = 2.55$$

Since the calculated z value of 2.55 falls between the z critical values −2.58 and 2.58, H_0 is not rejected. There is insufficient evidence to conclude that the true mean exam scores for the two groups differ.

P-value = 2P(z > 2.55) = 2(1 − .9946) = 2(.0054) = .0108. If a level of significance greater than .0108 had been used, then H_0 would have been rejected.

10.15 Let μ_1 denote the mean energy intake for elderly females and μ_2 the mean energy intake for elderly males.

H_0: $\mu_1 - \mu_2 = -300$ H_a: $\mu_1 - \mu_2 < -300$

$$z = \frac{(\bar{x}_1 - \bar{x}_2) - (-300)}{\sqrt{\dfrac{s_1^2}{n_1} + \dfrac{s_2^2}{n_2}}}$$

For $\alpha = 0.01$, reject H_0 if $z < -2.33$

$$z = \frac{(1395.4 - 1812.4) + 300}{\sqrt{(64.8)^2 + (95.4)^2}} = \frac{-117}{115.3265} = -1.01$$

Since the computed z is not less than −2.33, the null hypothesis is not rejected. The data does not support the conclusion that the mean energy intake for elderly males exceeds that of elderly females by more than 300 kcal/day.

10.17 The 99% confidence interval for $\mu_1 - \mu_2$ is

$$(\bar{x}_1 - \bar{x}_2) \pm 2.58\sqrt{\frac{s_1^2}{n_1} + \frac{s_2^2}{n_2}} \Rightarrow (33.40 - 35.39) \pm 2.58\sqrt{\frac{(.428)^2}{51} + \frac{(.294)^2}{54}}$$

$$\Rightarrow (-1.99) \pm 2.58\,(.0721) \Rightarrow -1.99 \pm .186 \Rightarrow (-2.176, -1.804)$$

It is not necessary to make any assumptions about the two salinity distributions. Even if they are not normal, the distribution of $\bar{x}_1 - \bar{x}_2$ will be approximately normal because of the large sample sizes.

Section 2

10.19 Let $\mu_1 - \mu_2$ denote the true difference in mean stopping distances using disk brakes and using pneumatic brakes for cars of this type.

H_0: $\mu_1 - \mu_2 = -10$ H_a: $\mu_1 - \mu_2 < -10$

$$t = \frac{(\bar{x}_1 - \bar{x}_2) - (-10)}{\sqrt{s_p^2\left(\dfrac{1}{n_1} + \dfrac{1}{n_2}\right)}} \quad \text{with d.f.} = 6 + 6 - 2 = 10$$

For $\alpha = 0.01$, reject H_o if $t < -2.76$.

$$s_p^2 = \left(\frac{5}{10}\right)(5.03)^2 + \left(\frac{5}{10}\right)(5.38)^2 = \left(\frac{1}{2}\right)(25.30) + \left(\frac{1}{2}\right)(28.94) = 27.12$$

$$t = \frac{(115.7 - 129.3) + 10}{\sqrt{27.12\left(\frac{1}{6} + \frac{1}{6}\right)}} = \frac{-3.6}{\sqrt{9.04}} = \frac{-3.6}{3.01} = -1.20$$

Since the calculated t of -1.20 is greater than -2.76, H_o is not rejected. There is insufficient evidence to conclude that the difference in mean stopping distances using disk brakes and pneumatic brakes is less than -10.

10.21 Let μ_1 denote the mean half-life of vitamin D in plasma for people on a high-fiber diet. Let μ_2 denote the mean half-life of vitamin D in plasma for people on a normal diet. Let $\mu_1 - \mu_2$ denote the true difference in mean half-life of vitamin D in plasma for people in these two groups (high-fiber minus normal).

H_o: $\mu_1 - \mu_2 = 0$ H_a: $\mu_1 - \mu_2 < 0$

$$t = \frac{(\bar{x}_1 - \bar{x}_2) - (0)}{\sqrt{s_p^2\left(\frac{1}{n_1} + \frac{1}{n_2}\right)}} \quad \text{with d.f.} = 6 + 7 - 2 = 11$$

For $\alpha = 0.01$, reject H_o if $t < -2.72$.

From the SAS output the value of the test statistic is $t = -2.9684$. Since the calculated t of -2.9684 is less than -2.72, H_o is rejected. There is sufficient evidence to conclude that the mean half-life of vitamin D is longer for those on a normal diet than for those on a high-fiber diet.

10.23 Let μ_1 denote the mean right leg strength for males and μ_2 the mean right leg strength for females.

$$s_p^2 = \frac{(13-1)(513)^2 + (14 - 1)(446)^2}{13 + 14 - 2} = 229{,}757.44$$

The 95% confidence interval for $\mu_1 - \mu_2$ based on this sample is $(2127 - 1843) \pm (2.06)\sqrt{229757.44\left(\frac{1}{12} + \frac{1}{13}\right)}$

$= 284 \pm 2.06(191.8856) = 284 \pm 395.28 = (-111.28, 679.28)$.

Based on this sample data, the difference between mean right leg strength of males and females is plausibly as much as 679.28 or as small as -111.28.

10.25 a) Let μ_1 denote the true average peak loudness for open-mouth chewing and μ_2 the true average peak loudness for closed mouth chewing. Then $\mu_1 - \mu_2$ denotes the difference between the means of open-mouthed and closed-mouth chewing.

$$s_p^2 = \frac{(10-1)(13)^2 + (10-1)(16)^2}{10 + 10 - 2} = \frac{9}{18}(169) + \frac{9}{18}(256)$$

$$s_p^2 = 84.5 + 128 = 212.5 \qquad s_p = \sqrt{212.5} = 14.5774$$

The 95% confidence interval for $\mu_1 - \mu_2$ based on this sample is

$$(63 - 54) \pm (2.10)\sqrt{212.5\left(\frac{1}{10} + \frac{1}{10}\right)} = 9 \pm 13.69 = (-4.69, 22.69).$$

This interval is rather wide because s_p^2 is large and both sample sizes are small. Observe that the interval includes 0, and so 0 is one of the plausible values of $\mu_1 - \mu_2$. That is, it is plausible that there is no difference in the mean loudness for open-mouth and closed-mouth chewing of potato chips.

b) Let μ_1 denote the true average peak loudness for closed-mouth chewing of potato chips and μ_2 the true average peak loudness for closed-mouth chewing of tortilla chips.

H_o: $\mu_1 - \mu_2 = 0$ $\quad H_a$: $\mu_1 - \mu_2 \neq 0$

$$t = \frac{(\bar{x}_1 - \bar{x}_2) - 0}{\sqrt{s_p^2\left(\frac{1}{n_1} + \frac{1}{n_2}\right)}} \qquad \text{with d.f.} = 10 + 10 - 2 = 18$$

For $\alpha = 0.01$, reject H_o if $t > 2.88$ or it $t < -2.88$

$$s_p^2 = \left(\frac{9}{18}\right)(16)^2 + \left(\frac{9}{18}\right)(16)^2 = 256$$

$$t = \frac{(54 - 53)}{\sqrt{256\left(\frac{1}{10} + \frac{1}{10}\right)}} = \frac{1.00}{7.155} = 0.1398$$

Since the calculated t value of .1398 is not in the rejection region, we fail to reject H_o. There is not sufficient evidence to conclude that there is a difference in the mean peak loudness for closed-mouth chewing of tortilla chips and potato chips.

c) Let μ_1 denote the true average peak loudness for fresh tortilla chips when chewing closed-mouth. Let μ_2 denote the true average peak loudness of stale tortilla chips when chewing closed-mouth.

H_0: $\mu_1 - \mu_2 = 0$ $\quad$ H_a: $\mu_1 - \mu_2 > 0$

$$t = \frac{(\bar{x}_1 - \bar{x}_2) - 0}{\sqrt{s_p^2\left(\frac{1}{n_1} + \frac{1}{n_2}\right)}} \quad \text{with d.f.} = 10 + 10 - 2 = 18$$

For $\alpha = 0.05$, reject H_0 if $t > 1.73$.

$$s_p^2 = \left(\frac{9}{18}\right)(14)^2 + \left(\frac{9}{18}\right)(16)^2 = 226$$

$$t = \frac{(56 - 53)}{\sqrt{226\left(\frac{1}{10} + \frac{1}{10}\right)}} = \frac{3.00}{6.7231} = 0.446$$

Since 0.446 is not in the rejection region, we fail to reject H_0. There is not sufficient evidence to conclude that there is a difference in the mean peak loudness when chewing fresh or stale tortilla chips closed-mouth.

10.27 Let μ_1 denote the mean number of imitations for infants who watch a human model. Let μ_2 denote the mean number of imitations for infants who watch a doll.

H_0: $\mu_1 - \mu_2 = 0$ $\quad$ H_a: $\mu_1 - \mu_2 > 0$

$$t = \frac{(\bar{x}_1 - \bar{x}_2) - 0}{\sqrt{s_p^2\left(\frac{1}{n_1} + \frac{1}{n_2}\right)}} \quad \text{with d.f.} = 12 + 15 - 2 = 25$$

For $\alpha = 0.01$, reject H_0 if the P-value is less than 0.01.

$$s_p^2 = \left(\frac{11}{25}\right)(1.6)^2 + \left(\frac{14}{25}\right)(1.3)^2 = 2.0728$$

$$t = \frac{(5.14 - 3.46)}{\sqrt{2.0728\left(\frac{1}{12} + \frac{1}{15}\right)}} = \frac{1.68}{.5576} = 3.01$$

P-value = $P(t > 3.01)$. From the table for critical t-values, the area under a t curve with 25 degrees of freedom to the right of 2.79 is .005, and the area to the right of 3.45 is .001. Hence, the P-value is less than .005, but greater than .001. Therefore, the P-value is less than .01 and so the null hypothesis is rejected. The data supports the conclusion that the mean number of imitations by infants who watch a human model is larger than the mean number of imitations by infants who watch a doll.

10.29 Let μ_1 and μ_2 denote the mean percent of body fat of women who participate and who do not participate regularly in aerobic exercise, respectively.

a) The 90% pooled t confidence interval for $\mu_1 - \mu_2$ is:

$$(\bar{x}_1 - \bar{x}_2) \pm 1.86\sqrt{s_p^2\left(\frac{1}{n_1} + \frac{1}{n_2}\right)}$$

$$s_p^2 = \frac{4(4.2)^2 + 4(3.4)^2}{8} = 14.6$$

The confidence interval then becomes

$$(20.1 + 20.4) \pm 1.86\sqrt{14.6\left(\frac{1}{5} + \frac{1}{5}\right)} = -0.3 \pm 1.86(2.417) = -0.3 \pm 4.495 = (-4.795, 4.195).$$

b) The interval obtained in (a) does contain zero. This means that 0 is one of the many plausible values for $\mu_1 - \mu_2$. Hence, it may be that there is no difference in the mean percent body fat of women who participate regularly in aerobic exercise and the mean percent body fat of women who do not participate in aerobic exercise.

c) The populations must both be normal with equal variances.

10.31 a)

$$s_p^2 = \left[\frac{6 - 1}{6 + 12 - 2}\right](78.4)^2 + \left[\frac{12 - 1}{6 + 12 - 2}\right](65.8)^2$$

$$= \left(\frac{5}{16}\right)(78.4)^2 + \left(\frac{11}{16}\right)(65.8)^2 = 4897.4275$$

b) Let μ_1 and μ_2 denote the true average yield when not under water stress and under water stress, respectively.

$H_0\colon \mu_1 - \mu_2 = 0 \quad H_a\colon \mu_1 - \mu_2 > 0$

$$t = \frac{(\bar{x}_1 - \bar{x}_2) - 0}{\sqrt{s_p^2\left(\frac{1}{n_1} + \frac{1}{n_2}\right)}} \quad \text{with d.f.} = 6 + 12 - 2 = 16$$

For $\alpha = 0.01$, reject H_0 if $t > 2.58$.

$$t = \frac{(376 - 234) - 0}{\sqrt{4897.4275\left(\frac{1}{6} + \frac{1}{12}\right)}} = \frac{142}{34.99} = 4.058$$

Since the calculated t of 4.058 exceeds the t critical 2.58, H_0 is rejected. The data does suggest that the true average yield under stress is less than that for the unstressed condition.

c) The P-value is less than .0005.

Section 3

10.33 a) If possible, treat each patient with both drugs. One drug used on one eye, the other drug used on the other eye. Then take observations (readings) of eye pressure on each eye. If this treatment method is not possible, then request the ophthalmologist to pair patients according to their eye pressure so that the two people in a pair have approximately equal eye pressure. Then treat one of the patients in the pair with the new drug and record the reduction in eye pressure. Treat the other person in that pair with the standard treatment and record the reduction in eye pressure. These two readings would constitute a pair. Repeat for each of the other pairs to obtain the paired sample data.

b) Both procedures above would result in paired data.

c) Select a group of persons to participate in the study. Randomly select a subset of this group to receive the new drug, and give the standard treatment to the remaining people. Measure reduction in eye pressure for both groups. The resulting observations would constitute independent samples.

10.35 Let μ_d be the mean difference between age at onset and age at diagnosis.

H_0: $\mu_d = -25$ H_a: $\mu_d < -25$

$$t = \frac{\bar{x}_d - (-25)}{\frac{s_d}{\sqrt{n}}} \quad \text{with d.f.} = 14$$

For $\alpha = 0.05$, reject H_0 if $t < -1.76$

$$t = \frac{-38.6 + 25}{\frac{23.18}{\sqrt{15}}} = \frac{-13.6}{5.985} = -2.27$$

Since −2.27 is less than −1.76, the null hypothesis is rejected. The data supports the conclusion that on average more than 25 months elapse between age at onset and age at diagnosis.

10.37 a) Let μ_d denote the true average difference in protein concentration between the two varieties of wheat (Sundance – Manitou).

H_o: $\mu_d = 0$ (no difference in average protein concentration)
H_a: $\mu_d \neq 0$ (average protein concentration for Sundance is not the same as average protein concentration of Manitou)

The test statistic is:

$$t = \frac{\bar{x}_d - 0}{\frac{s_d}{\sqrt{n}}} \quad \text{with d.f.} = 8$$

For $\alpha = .05$, reject H_o if $t < -2.31$ or if $t > 2.31$.

The differences are: –48, –27, –28, –26, –33, –12, –29, –27, –33.

From these: $\bar{x}_d = -29.222$ and $s_d = 9.351$

$$t = \frac{-29.222 - 0}{\frac{9.351}{\sqrt{9}}} = -9.37$$

Since the calculated t of –9.37 is less than the t critical value of –2.31, the null hypothesis is rejected. The data supports the conclusion that the mean protein concentration Sundance differs from the mean protein concentration of Manitou.

b) The 90% confidence interval for μ_d is:

$$\bar{d} \pm (\text{t critical}) \frac{s_d}{\sqrt{n}} = -29.222 \pm (1.86)\left(\frac{9.351}{\sqrt{9}}\right)$$

$$= 29.222 \pm 5.798 = (-35.020, -23.424)$$

With 90% confidence, it is estimated that the mean difference in protein concentration is between –35.020 and –23.424.

10.39 a) Let μ_d denote the mean change in blood lactate level for male racquetball players (After – Before).

The differences are: 5, 17, 23, 22, 17, 4, 18, 3.

From these, $\bar{x}_d = 13.625$ and $s_d = 8.2797$

The 95% confidence interval for μ_d is:

$$\bar{d} \pm (\text{t critical}) \frac{s_d}{\sqrt{n}} = 13.625 \pm (2.37)\left(\frac{8.2797}{\sqrt{8}}\right)$$

$= 13.625 \pm 6.938 = (6.687, 20.563)$

With 95% confidence, it is estimated that the mean change in blood lactate level for male racquetball players is between 6.687 and 20.563.

b) Let μ_d denote the mean change in blood lactate level for female racquetball players (After – Before).

The differences are: 10, 10, 6, 3, 0, 20, 7.

From these, $\bar{x}_d = 8.0$ and $s_d = 6.4031$

The 95% confidence interval for μ_d is:

$$\bar{d} \pm (\text{t critical}) \frac{s_d}{\sqrt{n}} = 8.0 \pm (2.45)\left(\frac{6.4031}{\sqrt{7}}\right)$$

$= 8.0 \pm 5.929 = (2.071, 13.929)$

With 95% confidence, it is estimated that the mean change in blood lactate level for female racquetball players is between 2.071 and 13.929.

c) Since the two intervals overlap (have values in common), this suggests that it is possible for the mean change for males and the mean change for females to have the same value.

10.41 a) Let μ_d denote the mean difference in blood pressure (dental setting minus medical setting).

H_o: $\mu_d = 0$ $\quad$ H_a: $\mu_d > 0$

The test statistic is:

$$t = \frac{\bar{x}_d - 0}{\frac{s_d}{\sqrt{n}}} \quad \text{with d.f.} = 59$$

For $\alpha = .01$, reject H_o if $t > 2.39$.

$$t = \frac{4.47 - 0}{\frac{8.77}{\sqrt{60}}} = 3.95$$

Since the calculated t of 3.95 exceeds the t critical value 2.39, H_o is rejected. Thus, the data does suggest that true mean blood pressure is higher in a dental setting than in a medical setting.

b) Let μ_d denote the true mean difference in pulse rate (dental minus medical).

H_o: $\mu_d = 0$ H_a: $\mu_d \neq 0$

The test statistic is:

$$t = \frac{\bar{x}_d - 0}{\frac{s_d}{\sqrt{n}}} \quad \text{with d.f.} = 59$$

For $\alpha = .05$, reject H_o if $t < -2.00$ or if $t > 2.00$.

$$t = \frac{-1.33 - 0}{\frac{8.84}{\sqrt{60}}} = -1.17$$

Since the t calculated value does not fall in the rejection region, H_o is not rejected. There is not sufficient evidence to conclude that mean pulse rates differ for a dental setting and a medical setting.

10.43 Let μ_d denote the true mean difference in improvement scores (experimental minus control).

H_o: $\mu_d = 0$ H_a: $\mu_d > 0$

The test statistic is:

$$t = \frac{\bar{x}_d - 0}{\frac{s_d}{\sqrt{n}}} \quad \text{with d.f.} = 6$$

For $\alpha = .10$, reject H_o if $t > 1.44$.

The differences are: −.3, −.1, .7, −.1, 1.1, −1.4, .2.

From these: $\bar{x}_d = 0.0143$ and $s_d = 0.7967$

$$t = \frac{.0143 - 0}{\frac{.7967}{\sqrt{7}}} = 0.047$$

Since the calculated t does not fall into the rejection region, H_o is not rejected. The experimental training method does not appear to be superior to the standard training method.

Section 4

10.45 Let π_1 denote the proportion of all spousal homicides resulting from shooting committed by a male and π_2 be the proportion of all spousal homicides accomplished by other means committed by a male.

H_o: $\pi_1 - \pi_2 = 0$ H_a: $\pi_1 - \pi_2 > 0$

$$z = \frac{(p_1 - p_2) - (0)}{\sqrt{\frac{p_c(1 - p_c)}{n_1} + \frac{p_c(1 - p_c)}{n_2}}}$$

For $\alpha = 0.01$, reject H_o if $z > 2.33$

$$p_1 = \frac{344}{429} = .8019 \qquad p_2 = \frac{468}{631} = .7417$$

$$p_c = \frac{344 + 468}{429 + 631} = \frac{812}{1060} = .7660$$

$$z = \frac{(.8019 - .7417) - 0}{\sqrt{.766(.234)\left(\frac{1}{429} + \frac{1}{631}\right)}} = \frac{.0602}{.0265} = 2.27$$

Since the computed z of 2.27 is not greater than 2.33, the null hypothesis is not rejected. The data does not support the conclusion that the proportion of spousal homicides resulting from shooting committed by males is greater than the proportion of spousal homicides accomplished by other means committed by males.

10.47 Let π_1 denote the proportion of returned surveys when a plain cover is used and π_2 denote the proportion of all returned surveys when a picture of a skydiver is used on the cover.

H_o: $\pi_1 - \pi_2 = 0$ H_a: $\pi_1 - \pi_2 < 0$

$$z = \frac{p_1 - p_2}{\sqrt{\frac{p_c(1 - p_c)}{n_1} + \frac{p_c(1 - p_c)}{n_2}}}$$

For $\alpha = .10$, reject H_o if $z < -1.28$.

$$p_1 = \frac{104}{207} = .5024 \text{ and } p_2 = \frac{109}{213} = .5117$$

$$p_c = \left(\frac{207}{420}\right)(.5024) + \left(\frac{213}{420}\right)(.5117) = .5071$$

$$z = \frac{(.5024 - .5117)}{\sqrt{\frac{.5071(.4929)}{207} + \frac{.5071(.4929)}{213}}} = \frac{-.0093}{.0448} = -.1906$$

Since –0.1906 is not less than –1.28, the null hypothesis is not rejected. The data do not support the researcher's claim that the response rate for the plain cover survey is lower than the response rate for the survey whose cover carried a picture of a skydiver.

10.49 Let π_1 denote the proportion of female Indian False Vampire bats that spend over five minutes in the air before locating food. Let π_2 denote the proportion of male Indian False Vampire bats that spend over five minutes in the air before locating food.

H_o: $\pi_1 - \pi_2 = 0$ H_a: $\pi_1 - \pi_2 \neq 0$

$$z = \frac{p_1 - p_2}{\sqrt{p_c(1 - p_c)\left[\frac{1}{n_1} + \frac{1}{n_2}\right]}}$$

For $\alpha = 0.01$, reject H_o if $z < -2.58$ or if $z > 2.58$.

$$p_1 = \frac{36}{193} = .1865, \quad p_2 = \frac{64}{168} = .3810,$$

$$p_c = \frac{(36 + 64)}{(193 + 168)} = \frac{100}{361} = .277$$

$$z = \frac{(.1865 - .3810)}{\sqrt{.277(.723)\left[\frac{1}{193} + \frac{1}{168}\right]}} = \frac{-.1945}{.0472} = -4.12$$

Since –4.12 is less than –2.58, H_o is rejected. There is sufficient evidence in the data to support the conclusion that the proportion of female Indian False Vampire bats who spend over five minutes in the air before locating food differs from that of male Indian False Vampire bats.

10.51 Let π_1 denote the true proportion who successfully complete parole for impulsive murderers and π_2 the true proportion who successfully complete parole for premeditated murderers.

$$(p_1 - p_2) \pm (\text{z critical})\sqrt{\frac{p_1(1 - p_1)}{n_1} + \frac{p_2(1 - p_2)}{n_2}}$$

$p_1 = .310, \quad p_2 = .550$

The 98% confidence interval for $\pi_1 - \pi_2$ is:

$$(.31 - .55) \pm (2.33)\sqrt{\frac{(.31)(.69)}{42} + \frac{(.55)(.45)}{40}}$$

$= (-.24) \pm (2.33)(.106) = -.24 \pm .247 = (-.487, .007)$

10.53 Let π_1 denote the proportion of women in 1970 who had never married and π_2 denote the proportion of women in 1990 who had never married.

H_0: $\pi_1 - \pi_2 = 0$ $\quad$ H_a: $\pi_1 - \pi_2 < 0$

$$z = \frac{p_1 - p_2}{\sqrt{p_c(1 - p_c)\left[\frac{1}{n_1} + \frac{1}{n_2}\right]}}$$

For $\alpha = 0.01$, reject H_0 if $z < -2.33$.

$$p_1 = .06,\ p_2 = .16 \text{ and } p_c = \frac{(12 + 32)}{200 + 200} = .11$$

$$z = \frac{(.06 - .16)}{\sqrt{.11(.89)\left[\frac{1}{200} + \frac{1}{200}\right]}} = \frac{-.10}{.0313} = -3.195$$

Since -3.195 is less than -2.33, the null hypothesis is rejected. This supports the conclusion that the proportion of women in 1990 who had never married exceeds the proportion of women in 1970 who had never married.

Supplementary Exercises

10.55 a) Let μ_1 denote the mean number of lightning flashes for single-peak storms and μ_2 the mean number of lightning flashes for multiple-peak storms.

H_0: $\mu_1 - \mu_2 = 0$ $\quad$ H_a: $\mu_1 - \mu_2 \neq 0$

The test statistic is:

$$t = \frac{(\bar{x}_1 - \bar{x}_2) - 0}{\sqrt{s_p^2\left(\frac{1}{n_1} + \frac{1}{n_2}\right)}} \quad \text{with d.f.} = 7 + 4 - 2 = 9.$$

For $\alpha = 0.05$, reject H_0 if $t < -2.26$ or if $t > 2.26$.

From the data:

$\bar{x}_1 = 66.143,\ s_1^2 = 1007.81,\ \bar{x}_2 = 274.5,\ s_2^2 = 11104.33$

$$s_p^2 = \frac{(6)(1007.81)}{9} + \frac{(3)(11104.33)}{9} = 4373.32$$

$$t = \frac{(66.143 - 274.5)}{\sqrt{4373.32\left(\frac{1}{7} + \frac{1}{4}\right)}} = \frac{-208.357}{41.45} = -5.03$$

Since the calculated t value of –5.03 falls in the rejection region, the null hypothesis is rejected. The data suggests that there is a difference between the mean number of lightening flashes for single-peak and multiple-peak storms.

b) The distributions of the number of lightning flashes of single-peak storms and multiple-peak storms are normal. The population variances of these two distributions are equal.

Comment: The fact that there is such a large difference between the two sample variances suggests that perhaps the spreads of the two populations are not equal. If this is the case, then the pooled t test should not be used.

10.57 Let μ_1 denote the mean ratio for young men and μ_2 the mean ratio for elderly men.

H_0: $\mu_1 - \mu_2 = 0$ $\quad H_a$: $\mu_1 - \mu_2 > 0$

If the populations of ratios are normally distributed with equal variances, then an appropriate test statistic is:

$$t = \frac{(\bar{x}_1 - \bar{x}_2) - 0}{\sqrt{s_p^2\left(\frac{1}{n_1} + \frac{1}{n_2}\right)}} \quad \text{with d.f.} = 13 + 12 - 2 = 23$$

For $\alpha = 0.05$, reject H_0 if $t > 1.71$

$s_1^2 = 13(.22)^2 = .6292$

$s_2^2 = 12(.28)^2 = .9408$

$$s_p^2 = \frac{12(.6292) + 11(.9408)}{23} = .7782$$

$$t = \frac{(7.47 - 6.71) - 0}{\sqrt{.7782\left(\frac{1}{13} + \frac{1}{12}\right)}} = \frac{0.76}{.3531} = 2.15$$

Since the calculated t of 2.15 exceeds the critical t of 1.71, the null hypothesis is rejected. The data supports the conclusion that the mean ratio for young men exceeds that for elderly men.

10.59 Let μ_1 denote the mean score of male faculty on the MSQ and μ_2 the mean score of female faculty on the MSQ.

H_o: $\mu_1 - \mu_2 = 0$ $\quad$ H_a: $\mu_1 - \mu_2 \neq 0$

The test statistic is: $z = \dfrac{(\bar{x}_1 - \bar{x}_2) - 0}{\sqrt{\dfrac{s_1^2}{n_1} + \dfrac{s_2^2}{n_2}}}$.

For $\alpha = 0.01$, reject H_o if $z < -2.58$ or if $z > 2.58$.

$$z = \frac{(75.43 - 72.54)}{\sqrt{\frac{(10.53)^2}{115} + \frac{(13.08)^2}{105}}} = 1.79$$

Since the z calculated value of 1.79 does not fall in the rejection region, the null hypothesis is not rejected. The data does not suggest that male and female academic employees differ with respect to mean score on the MSQ.

10.61 Let μ_1 denote the true mean age at death for female SIDS victims and μ_2 the true mean age at death for male SIDS victims.

From the data,

$\bar{x} = 103.6$, $s_1^2 = 2154.3$, $\bar{x}_2 = 89.7$, $s_2^2 = 1638.238$

$$s_p^2 = \left(\frac{4}{10}\right)(2154.3) + \left(\frac{6}{10}\right)(1638.238) = 1844.66$$

The confidence interval is:

$$(103.6 - 89.7) \pm (2.23)\sqrt{\frac{1844.66}{5} + \frac{1844.66}{7}}$$

$(13.9) \pm (2.23)(25.149) = 13.9 \pm 56.08 = (-42.18, 69.98)$

With 95% confidence, it is estimated that $\mu_1 - \mu_2$ is between −42.18 and 69.98. Since this interval contains zero, the analysis supports the conclusion that there may be no difference between the true mean ages at death of male and female SIDS victims.

10.63 For each of the tests, let μ_1 denote the mean blood characteristic for the mice exposed to an electric field and μ_2 the mean blood characteristic for the control group.

H_0: $\mu_1 - \mu_2 = 0$ $\quad H_a$: $\mu_1 - \mu_2 \neq 0$

The test statistic is: $z = \dfrac{(\bar{x}_1 - \bar{x}_2) - 0}{\sqrt{\dfrac{s_1^2}{n_1} + \dfrac{s_2^2}{n_2}}}$.

For $\alpha = 0.05$, reject H_0 if $z < -1.96$ or if $z > 1.96$.

For glucose:

$$z = \frac{(139.20 - 136.30)}{\sqrt{\dfrac{(16.1)^2}{45} + \dfrac{(12.7)^2}{45}}} = \frac{2.90}{3.06} = .95$$

Since the calculated z value of .95 does not fall in the rejection region, the null hypothesis is not rejected. The data suggests that there is no difference in mean glucose levels of the two populations.

For potassium:

$$z = \frac{(7.62 - 7.44)}{\sqrt{\dfrac{(.54)^2}{45} + \dfrac{(.53)^2}{45}}} = \frac{0.180}{0.113} = 1.60$$

Since the z calculated value of 1.6 does not fall in the rejection region, the null hypothesis is not rejected. The data suggests that there is no difference in mean potassium levels for the two populations.

For total protein:

$$z = \frac{(6.61 - 6.63)}{\sqrt{\dfrac{(.34)^2}{45} + \dfrac{(.27)^2}{45}}} = \frac{-.02}{.065} = -.31$$

Since the z calculated value of –.31 does not fall in the rejection region, the null hypothesis is not rejected. The data suggests that there is no difference in mean total protein levels for the two populations.

For cholesterol:

$$z = \frac{(67.8 - 69.0)}{\sqrt{\dfrac{(9.38)^2}{45} + \dfrac{(11.4)^2}{45}}} = \frac{-1.2}{2.2} = -0.55$$

Since the z calculated value of −0.55 does not fall in the rejection region, the null hypothesis is not rejected. The data suggests that there is no difference in mean cholesterol levels for the two populations.

10.65 Let π_1 denote the true proportion of adults born deaf who remove the implants. Let π_2 denote the true proportion of adults who went deaf after learning to speak who remove the implants.

H_0: $\pi_1 - \pi_2 = 0$ H_a: $\pi_1 - \pi_2 \neq 0$

The test statistic is: $z = \dfrac{(p_1 - p_2)}{\sqrt{\dfrac{p_c(1-p_c)}{n_1} + \dfrac{p_c(1-p_c)}{n_2}}}$.

For $\alpha = 0.01$, reject H_0 if $z < -2.58$ or if $z > 2.58$.

From the data, $p_1 = \dfrac{75}{250} = .3$, $p_2 = \dfrac{25}{250} = .1$,

$$p_c = \left(\frac{250}{500}\right)(.3) + \left(\frac{250}{500}\right)(.1) = .2$$

$$z = \frac{.3 - .1}{\sqrt{\dfrac{(.28)(.8)}{250} + \dfrac{(.2)(.8)}{250}}} = \frac{.2}{.03577} = 5.59$$

Since the calculated z value of 5.59 falls in the rejection region, the null hypothesis is rejected. The data does support the fact that the true proportion who remove the implants differs in those that were born deaf from that of those who went deaf after learning to speak.

10.67

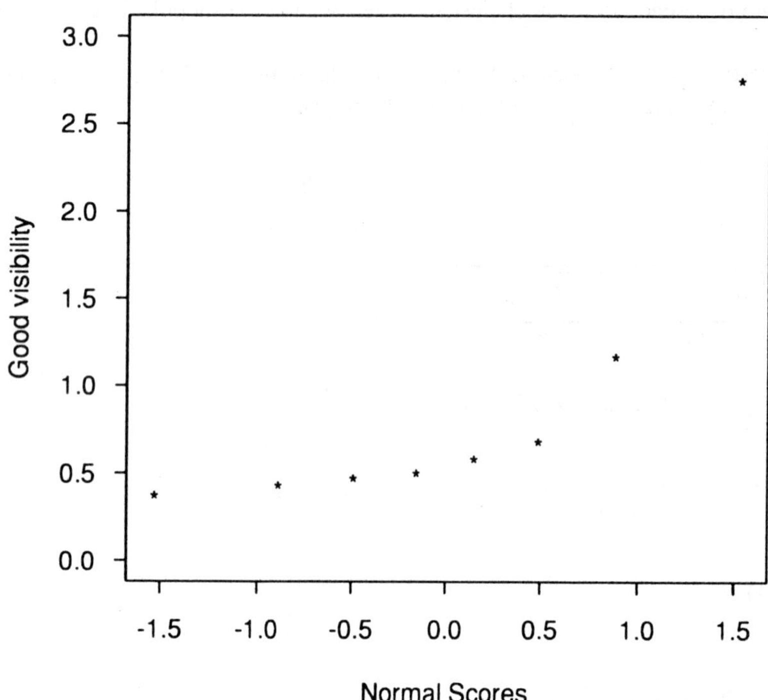

The normal probability plot for good visibility has pronounced curvature in it. This strongly suggests that the distribution from which the good visibility sample was taken is not distributed like that of a normal.

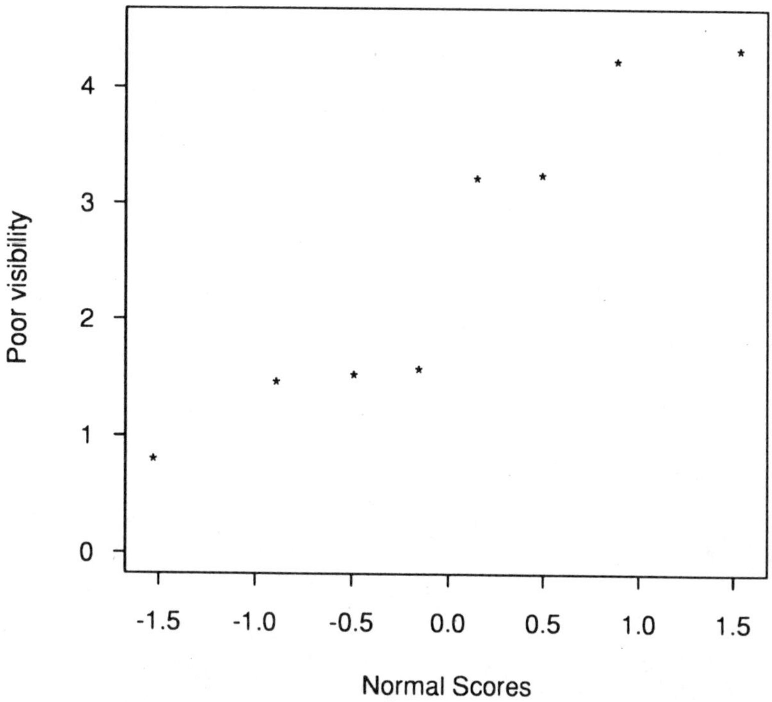

The normal probability plot for poor visibility is reasonably straight. This suggests that the distribution from which the poor visibility sample was taken is distributed like that of a normal. Since the pooled t-test requires that both populations being sampled be normally distributed, the use of the pooled t-test does not appear to be justified.

Chapter 11
Regression and Correlation: Descriptive Methods

Section 1

11.1 a)

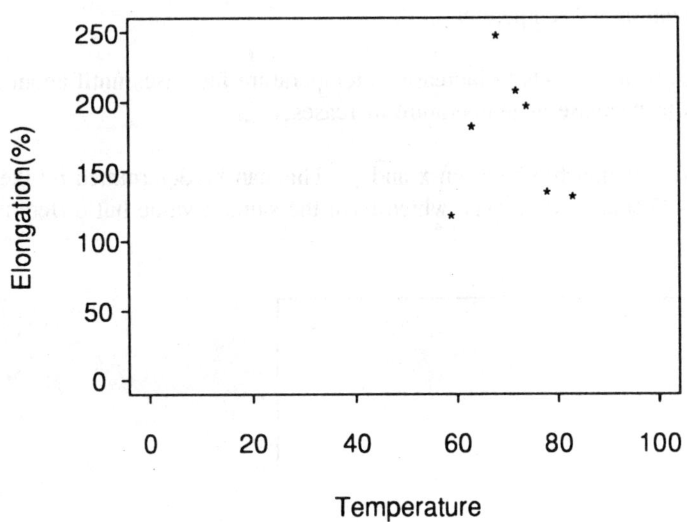

b)

Elongation(%)

250 225 200 175 150 125 100

55 60 65 70 75 80 85

Temperature

This plot is preferable to the plot in part (a), because it reveals more clearly the possible curvilinear relationship between elongation and temperature.

c) The graph suggests that elongation tends to increase as temperature increases until about seventy degrees and then elongation tends to decrease as temperature increases.

11.3 a) There is not a deterministic relationship between x and y. This can be determined by the fact that there are two data points, (100, 222) and (100, 241), which have the same x-value but different y-values.

b)

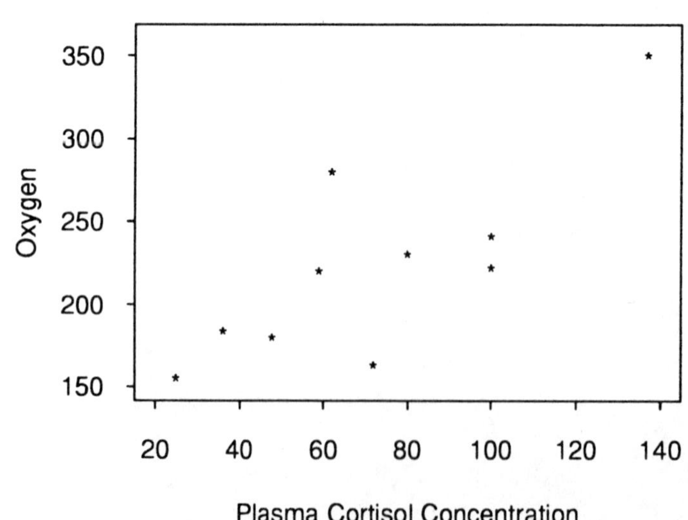

c) There appears to be a tendency for oxygen consumption rate to increase as plasma cortisol concentration increases.

11.5 There are several observations that have identical or nearly identical x-values yet different y-values. Therefore, the value of y is not determined solely by x, but also by various other factors. There appears to be a general tendency for y to decrease in value as x increases in value. There are two data points which are far removed from the remaining data points. These two data points have large x values and small y values. Their presence might have an undue influence on a line fit to the data.

11.7 a)

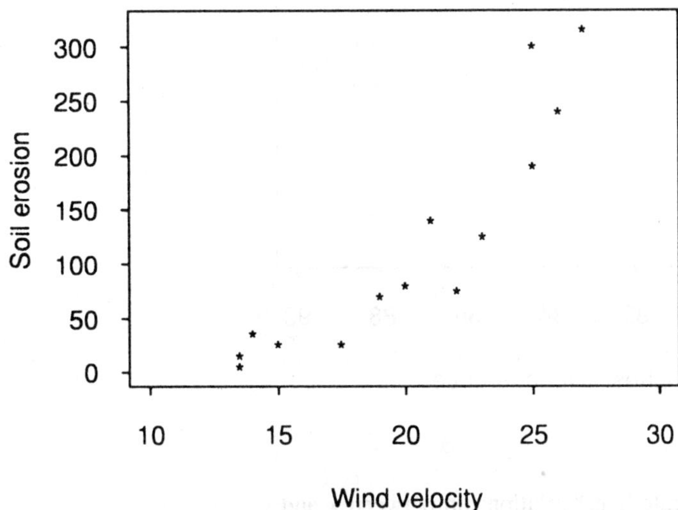

There appears to be a strong relationship between x and y. The relationship appears not to be linear in nature, but of a curved pattern (perhaps parabolic).

b)

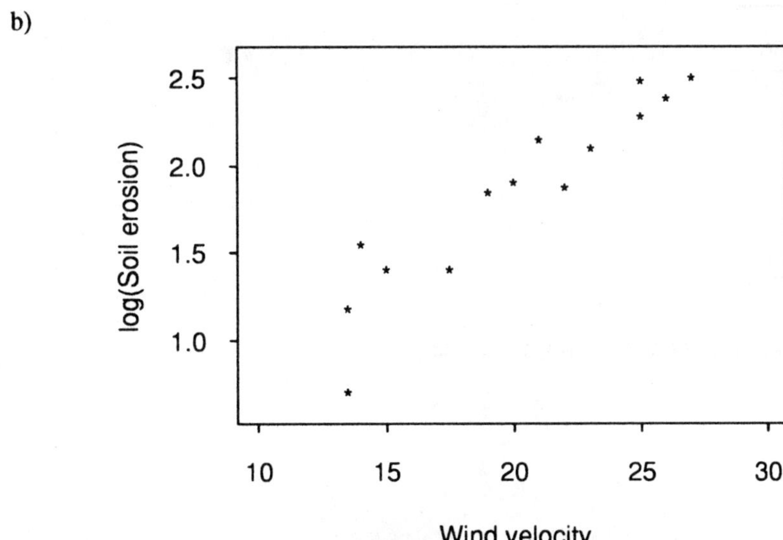

The plot reveals a strong linear relationship between x and log (y).

Section 2

11.9 a)

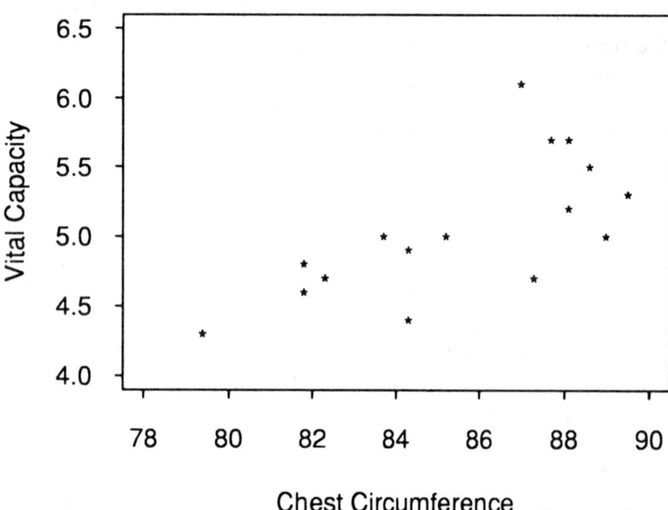

The graph reveals a moderate linear relationship between x and y.

b)

$$b = \frac{6933.48 - \frac{(1368.1)(80.9)}{16}}{117123.85 - \frac{(1368.1)^2}{16}}$$

$$b = \frac{6933.48 - 6917.456}{117123.85 - 116981.101}$$

$$b = \frac{16.024}{142.749} = 0.1123$$

$$a = \frac{80.9}{16} - 0.1123\left(\frac{1368.1}{16}\right)$$

$a = 5.0563 - 0.1123(85.5063) = 5.0563 - 9.6024 = -4.546$

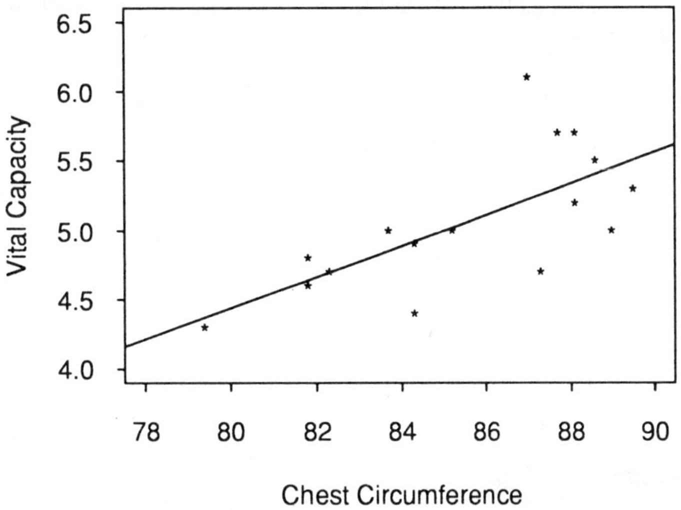

c) The change in vital capacity associated with a 1 cm. increase in chest circumference is .1123.

The change in vital capacity associated with a 10 cm. increase in chest circumference is 10(.1123) = 1.123

d) $\hat{y} = -4.54 + .1123(85) = 5.0055$

e) No; this is shown by the fact that there are two data points in the data set whose x-values are 81.8, but these data points have different y-values.

11.11 a) There are two data points, (12,12) and (12,23) which have the same x-value but different y-values.

b)

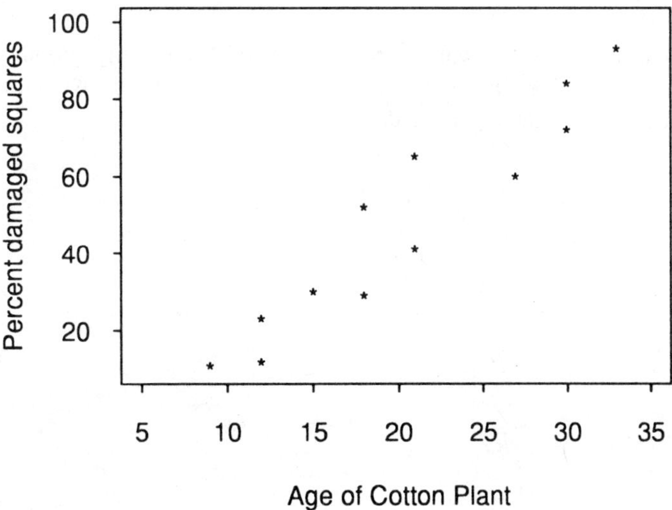

The scatter plot suggests a fairly strong linear relationship between x and y, with a fairly tight clustering about a visually fitted line. This suggests that the least squares line will effectively summarize the relationship between the two variables.

c) $\Sigma xy - \dfrac{(\Sigma x)(\Sigma y)}{n} = 14022 - \dfrac{(246)(572)}{12} = 2296$

$\Sigma x^2 - \dfrac{(\Sigma x)^2}{n} = 5742 - \dfrac{(246)^2}{12} = 699$

$b = \dfrac{2296}{699} = 3.2847$

$a = 47.667 - 3.2847(20.5) = -19.67$

The equation of the least squares line is

$\hat{y} = -19.67 + 3.2847x$

d) When $x = 20$, $\hat{y} = -19.67 + 3.2847(20) = 46.02$

11.13 a) Using the column of the MINITAB output labeled *Coef*, the value of the intercept of the least squares lines is 4027 and the value of the slope of the least squares line is −577.9. The equation of the least squares line is

$\hat{y} = 4027 - 577.9x$

b) The b value of −577.9 is the estimate of the change in myoglobin level associated with a one unit increase in finishing time.

c) $\hat{y} = 4027 - 577.9(8) = -596.2$

The least squares equation yields a negative value for the estimated level of myoglobin when the finishing time is 8h. This is clearly unreasonable since myoglobin level cannot be a negative value. This occurs because $x = 8$ is outside the range of the available data.

11.15 a)

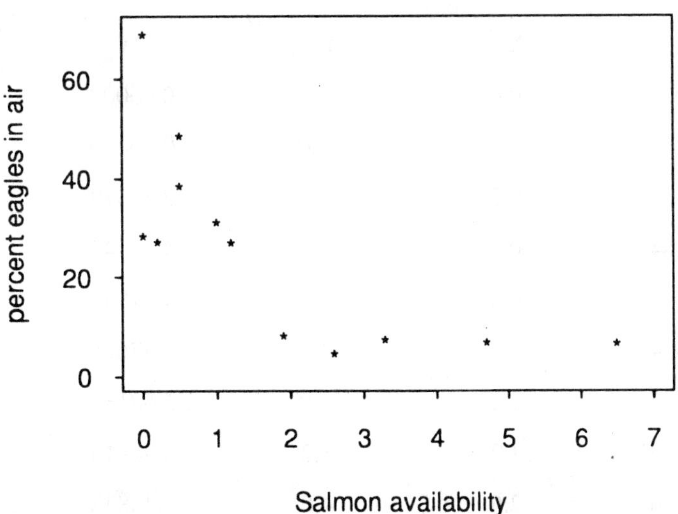

The plot is curvilinear, not linear. One would not feel comfortable summarizing the relationship between a and y by reporting the equation of the least squares line.

b)

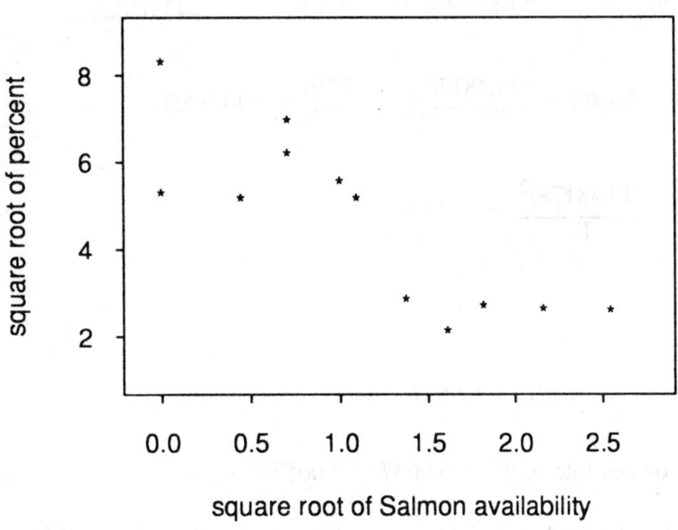

This relationship between the square root of x and the square root of y appears to be approximately linear.

c)

X'	Y'	X'²	Y'²	X'Y'
0.00000	5.31037	0.0	28.2	0.00000
0.00000	8.30662	0.0	69.0	0.00000
0.44721	5.19615	0.2	27.0	2.32379
0.70711	6.20484	0.5	38.5	4.38748
0.70711	6.95701	0.5	48.4	4.91935
1.00000	5.57674	1.0	31.1	5.57674
1.09545	5.18652	1.2	26.9	5.68155
1.37840	2.86356	1.9	8.2	3.94715
1.61245	2.14476	2.6	4.6	3.45832
1.81659	2.72029	3.3	7.4	4.94166
2.16795	2.64575	4.7	7.0	5.73585
2.54951	2.60768	6.5	6.8	6.64831
13.48178	55.72029	22.4	303.1	47.62020

$$\Sigma x'y' - \frac{(\Sigma x')(\Sigma y')}{n} = 47.6202 - \frac{(13.48178)(55.72029)}{12} = -14.98052$$

$$\Sigma x'^2 - \frac{(\Sigma x')^2}{n} = 22.4 - \frac{(13.48178)^2}{12} = 7.25347$$

$$b = \frac{-14.98052}{7.25347} = -2.0653$$

a = [55.72029 – (–2.0653)(13.48178)]/12 = 6.9637

The equation of the least squares line is $\hat{y}' = 6.9637 - 2.0653x'$.

When x = 2.0, x' = 1.4142. The predicted y' for x' = 1.4142 is y' = 6.9637 – 2.0653(1.4142) = 4.0429.

The predicted value for y when x = 2 would be $(4.0428)^2 = 16.3442$.

Section 3

11.17 a)

x	y	predicted	residual
15	23	40.4048	−17.4048
19	52	42.6245	9.3755
31	65	49.2837	15.7163
39	55	53.7231	1.2769
41	32	54.8330	−22.8330
44	60	56.4978	3.5022
47	78	58.1626	19.8374
48	59	58.7175	0.2825
55	61	62.6020	−1.6020
65	60	68.1513	−8.1513

b)

$$\begin{aligned}\text{SSResid} &= (-17.4048)^2 + (9.3755)^2 + \ldots + (-1.6020)^2 + (-8.1513)^2 \\ &= 302.9271 + 87.9000 + \ldots + 2.5664 + 66.4437 \\ &= 1635.6833\end{aligned}$$

c)

$$\text{SSTo} = 31993 - \frac{(545)^2}{10} = 31993 - 29702.5 = 2290.50$$

$$r^2 = 1 - \frac{1635.6833}{2290.5000} = 1 - .7141 = .2859$$

The least squares line does not give very accurate predictions. Only 28.59% of the observed variation in age is explained by the linear relationship between percent of root with transparent dentine and age.

11.19 a)

$$\begin{aligned}\Sigma x^2 - \frac{(\Sigma x)^2}{n} &= 62.600235 - \frac{(22.027)^2}{12} \\ &= 62.600235 - 40.43239 \\ &= 22.16784\end{aligned}$$

$$\begin{aligned}\Sigma xy - \frac{(\Sigma x)(\Sigma y)}{n} &= 1114.495 - \frac{(22.027)(793)}{12} \\ &= 1114.495 - 1455.61758 \\ &= -341.122583\end{aligned}$$

$$b = \frac{-341.122583}{22.16784} = -15.38817$$

$$\begin{aligned}a &= 66.08333 - (-15.38817)(1.83558) \\ &= 66.08333 + 28.24622 \\ &= 94.32955\end{aligned}$$

The least squares equation is $\hat{y} = 94.33 - 15.388x$

b) $$SSTo = \Sigma y^2 - \frac{(\Sigma y)^2}{n} = 57939 - \frac{(793)^2}{12}$$

$= 57939 - 52404.08333$

$= 5,534.91667$

SSResid = 57939 − 94.32955(793) − (−15.38817)(1114.495)

= 57939 − 74803.33315 + 17150.03852

= 285.70537

c) $$r^2 = 1 - \frac{285.70537}{5534.91667} = 1 - .05162 = .94838 \text{ or } 94.838\%$$

11.21 $$r^2 = 1 - \frac{1235.470}{25321.368} = 0.9512$$

The coefficient of determination reveals that 95.12% of the total variation in hardness of molded plastic can be explained by the linear relationship between hardness and the amount of time elapsed since termination of the molding process.

11.23 a,b) The computations were done using the least squares equation $\hat{y} = 61.536984 - 3.406909x$.

ROW	X	Y	PRED Y	RESIDUAL	RESID SQ
1	−5.0	43	44.5024	−1.5024	0.257
2	−3.2	50	50.6349	−0.6349	0.403
3	−2.2	61	54.0418	6.9582	48.417
4	−1.7	63	55.7452	7.2548	52.632
5	−1.6	47	56.0859	−9.0859	82.554
6	−1.5	57	56.4266	0.5734	0.329
7	−0.9	51	58.4708	−7.4708	55.812
8	0.0	60	61.5370	−1.5370	2.362
9	0.0	67	61.5370	5.4630	29.845
10	1.2	76	65.6253	10.3747	107.635
11	1.6	70	66.9880	3.0120	9.072
12	1.7	51	67.3287	−16.3287	266.627
13	2.8	74	71.0763	2.9237	8.548

SSRESID = 666.493

c) SSResid = 46940 − (61.536984)(770) − (3.406909)(−325.8) = 666.4933

d) SSResid = 46940 − (61.5)(770) − (3.4)(−325.8) = 692.72

Yes, rounding has made a substantial difference.

e) $$r^2 = 1 - \frac{666.4937}{1332.3077} = 0.4997$$

Only 49.97% of the total variation in cholesterol level can be explained using a straight line model that relates cholesterol level to percent change in body weight. Looking at the graph of Example 11.5, the relationship seems approximately linear. The data point (1.7, 51) may be part of the reason for a rather low r^2 value.

Section 4

11.25 The statement is incorrect. The correlation coefficient measures the extent to which x and y are linearly related. They may have some other strong relationship and yet have a correlation of zero.

11.27 $$r = \frac{1410933.1}{\sqrt{(557582.1)(6347190)}} = \frac{1410933.1}{1881244.144} = 0.75$$

Since r = .75, the correlation between dry weight and volume would be described as a moderate positive correlation.

11.29 $n = 16$, $\Sigma x = 160.5$, $\Sigma x^2 = 1997.19$, $\Sigma y = 53.9$, $\Sigma y^2 = 321.55$, $\Sigma xy = 623.35$

$$\Sigma x^2 - \frac{(\Sigma x)^2}{n} = 1997.19 - \frac{(160.5)^2}{16} = 387.174$$

$$\Sigma y^2 - \frac{(\Sigma y)^2}{n} = 321.55 - \frac{(53.9)^2}{16} = 139.974$$

$$\Sigma xy - \frac{(\Sigma x)(\Sigma y)}{n} = 623.35 - \frac{(160.5)(53.9)}{16} = 82.666$$

$$r = \frac{82.666}{\sqrt{(387.174)(139.974)}} = .355$$

Since r = .355, the correlation between carbon monoxide concentration and benzo(a)pyrene concentration would be described as a weak positive correlation. Therefore, the relationship cannot be very accurately modeled by a straight line.

11.31 a)

$$\Sigma x^2 - \frac{(\Sigma x)^2}{n} = 251970 - \frac{(1950)^2}{18} = 40720$$

$$\Sigma y^2 - \frac{(\Sigma y)^2}{n} = 130.6074 - \frac{(47.92)^2}{18} = 3.0337$$

$$\Sigma xy - \frac{(\Sigma x)(\Sigma y)}{n} = 5530.92 - \frac{(1950)(47.92)}{18}$$

$$= 339.5867$$

$$r = \frac{339.5867}{\sqrt{(40720)(3.0337)}} = .966$$

b) The percentage of the observed variation in percent dry weight explained by the approximate linear relationship is

$r^2 = (.966)^2 = .9335$, which converts to 93.35%.

11.33 a)

$$\Sigma xy - \frac{(\Sigma x)(\Sigma y)}{n} = 86264 - \frac{(497)(1219)}{7} = 86264 - 86549 = -285$$

$$\Sigma x^2 - \frac{(\Sigma x)^2}{n} = 35707 - \frac{(497)^2}{7} = 35707 - 35287 = 420$$

$$\Sigma y^2 - \frac{(\Sigma y)^2}{n} = 225779 - \frac{(1219)^2}{7} = 225779 - 212280.1429 = 13498.8571$$

$$r = \frac{-285}{\sqrt{(420)(13498.8571)}} = \frac{-285}{2381.0754} = -0.1197$$

b) Since r = –.1197, which is relatively close to 0, it does not appear that there is a strong linear relationship between the temperature of the Mozzarella cheese and elongation at failure.

c) The plot in Exercise 11.1 suggests a strong non-linear (perhaps quadratic) relationship.

11.35 The sample correlation coefficient would be closest to –.9. This is because there is an almost perfect inverse relationship between rate and time required to travel a fixed distance. That is, as rate increases, time required to traverse the fixed distance decreases.

11.37 The value of the sample correlation coefficient using the squared y values would not necessarily be approximately one. If the y values are greater than 1, then the squared y values would differ from each other by more than the y values differ from one another. Hence, the relationship between x and y^2 would be less like a straight line, and the resulting value of the correlation coefficient would decrease.

11.39 a) $\Sigma x = 561$, $\Sigma x^2 = 37695$, $\Sigma xy = 38281$, $\Sigma y^2 = 40223$, $\Sigma y = 589$

$$\Sigma x^2 - \frac{(\Sigma x)^2}{n} = 37695 - \frac{(561)^2}{9} = 2726$$

$$\Sigma y^2 - \frac{(\Sigma y)^2}{n} = 40223 - \frac{(589)^2}{9} = 1676.222$$

$$\Sigma xy - \frac{(\Sigma x)(\Sigma y)}{n} = 38281 - \frac{(561)(589)}{9} = 1566.667$$

$$r = \frac{1566.667}{\sqrt{(2726)(1676.22)}} = .733$$

b) The $(\bar{x},\bar{y})$ pairs are (78.67, 76), (65,68) and (43.33, 52.33). Ignoring the bar notation

$\Sigma x = 187$, $\Sigma x^2 = 12291.4578$, $\Sigma xy = 12666.3789$, $\Sigma y^2 = 13138.4289$, $\Sigma y = 196.33$

$$\Sigma x^2 - \frac{(\Sigma x)^2}{n} = 12291.4578 - \frac{(187)^2}{3} = 635.1245$$

$$\Sigma y^2 - \frac{(\Sigma y)^2}{n} = 13138.4289 - \frac{(196.33)^2}{3} = 289.9393$$

$$\Sigma xy - \frac{(\Sigma x)(\Sigma y)}{n} = 12666.3789 - \frac{(187)(196.33)}{3}$$

$$= 428.475567$$

$$r = \frac{428.475567}{\sqrt{(635.1245)(289.9393)}} = .9985$$

c)

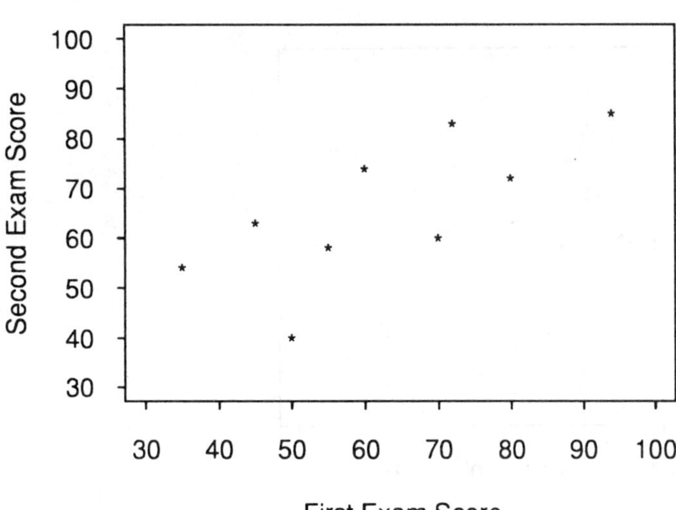

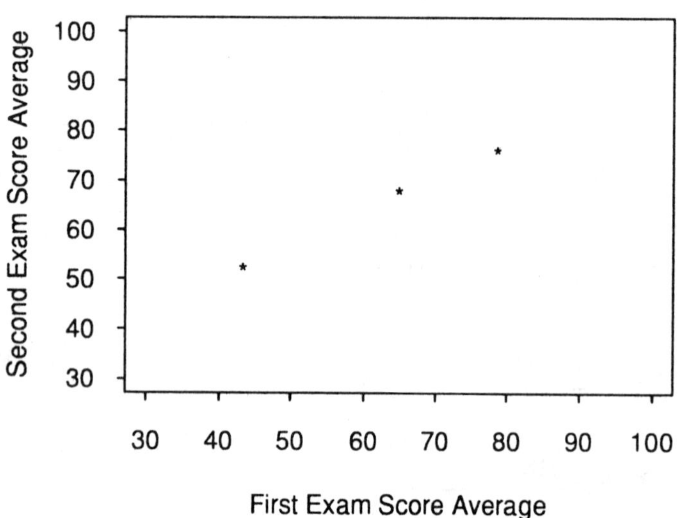

Computing means yields a typical value for the data, and the resulting group of means has less variability than the original values. Hence, the correlation will be larger when based on averages than when based on the original values.

Supplementary Exercises

11.41 a)

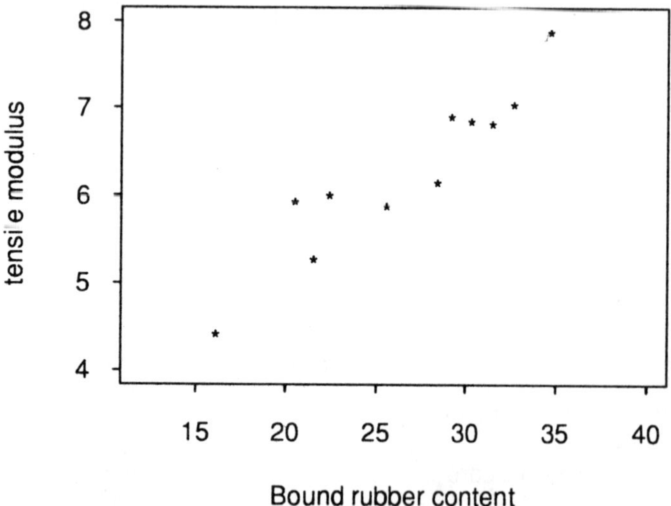

The scatterplot above suggests the plausibility of a linear relationship between bound rubber content and tensile modulus.

b) From the MINITAB output, the value of a = 2.2249 and the value of b = 0.15212.

The equation of the estimated regression line is

$$\hat{y} = 2.2249 + .15212x$$

The value of b indicates that when bound rubber content increases one unit, the average increase in tensile modulus increase by .15212 units.

The predicted value for tensile modulus when bound rubber content is 25, is

$$\hat{y} = 2.2249 + .15212(25) = 6.0279$$

c) From the MINITAB output, SSResid = SSE = 1.0673 and SSTo = 8.9957. From these,

$r^2 = 1 - \dfrac{1.0673}{8.9957} = 1 - .1186 = .8814$, which is 88.14%. Therefore, 88.14 % of the observed variation in tensile modulus can be explained by the approximate linear relationship between bound rubber content and tensile modulus.

d) The value of the sample correlation coefficient is

$$\sqrt{.8814} = .9388.$$

11.43 a) y = stride rate x = speed

The summary values are: $n = 11$, $\Sigma x = 205.4$, $\Sigma x^2 = 3880.08$, $\Sigma y = 35.16$, $\Sigma y^2 = 112.681$, $\Sigma xy = 660.13$

$$\bar{x} = \frac{205.4}{11} = 18.6727, \quad \bar{y} = \frac{35.16}{11} = 3.1964$$

$$\Sigma xy - \frac{(\Sigma x)(\Sigma y)}{n} = 660.13 - \frac{(205.4)(35.16)}{11} = 3.5969$$

$$\Sigma x^2 - \frac{(\Sigma x)^2}{n} = 3880.08 - \frac{(205.4)^2}{11} = 44.7018$$

$$\Sigma y^2 - \frac{(\Sigma y)^2}{n} = 112.681 - \frac{(35.16)^2}{11} = .2969$$

$$b = \frac{3.5969}{44.7018} = .0805$$

$$a = 3.1964 - (.0805)(18.6727) = 1.6932$$

The equation of the least squares line for predicting stride rate from speed is

$$\hat{y} = 1.6932 + .0805x$$

b) x = stride rate y = speed

The summary values are: $n = 11$, $\Sigma x = 35.16$, $\Sigma x^2 = 112.681$, $\Sigma y = 205.4$, $\Sigma y^2 = 3880.08$, $\Sigma xy = 660.13$

$$\bar{x} = \frac{35.16}{11} = 3.1964, \quad \bar{y} = \frac{205.4}{11} = 18.6727,$$

$$\Sigma xy - \frac{(\Sigma x)(\Sigma y)}{n} = 660.13 - \frac{(35.16)(205.4)}{11} = 3.5969$$

$$\Sigma x^2 - \frac{(\Sigma x)^2}{n} = 112.681 - \frac{(35.16)^2}{11} = .2969$$

$$\Sigma y^2 - \frac{(\Sigma y)^2}{n} = 3880.08 - \frac{(205.4)^2}{11} = 44.7018$$

$$b = \frac{3.5969}{.2969} = 12.1149$$

$$a = 18.6727 - (12.1149)(3.1964) = -20.0514$$

The equation of the least squares line for predicting speed from stride rate is

$$\hat{y} = -20.0514 + 12.1149x$$

c) The coefficient of determination for the "stride rate on speed" regression is

$$r^2 = \frac{(3.5969)^2}{(44.7018)(.2969)} = .9748$$

The coefficient of determination for the "speed on stride rate" regression is

$$r^2 = \frac{(3.5969)^2}{(.2969)(44.7018)} = .9748$$

The relationship between the two r^2's is that they are equal in value.

11.45

ROW	X	Y	PRED Y	RESIDUAL	
1	1.03	57.6	59.5658	-1.9658	
2	1.43	60.1	53.3458	6.7542	
3	1.79	45.7	47.7477	-2.0477	
4	1.74	35.6	48.5252	-12.9252	*
5	0.95	68.4	60.8098	7.5902	
6	1.47	41.5	52.7237	-11.2237	*
7	0.54	75.0	67.1853	7.8147	
8	1.76	51.3	48.2142	3.0858	
9	1.76	50.0	48.2142	1.7858	
10	1.09	60.5	58.6328	1.8672	
11	1.97	57.0	44.9487	12.0513	*
12	1.65	44.0	49.9247	-5.9247	
13	1.69	52.3	49.3027	2.9973	
14	1.41	40.7	53.6568	-12.9568	*
15	1.68	58.4	49.4582	8.9418	
16	1.26	43.8	55.9893	-12.1893	*
17	1.70	48.8	49.1472	-0.3472	
18	1.92	52.9	45.7262	7.1738	
19	1.25	48.6	56.1448	-7.5448	
20	1.71	53.1	48.9917	4.1083	
21	1.22	51.9	56.6113	-4.7113	
22	1.24	56.0	56.3003	-0.3003	
23	1.44	51.9	53.1903	-1.2903	
24	1.62	57.6	50.3912	7.2088	
25	1.47	50.0	52.7237	-2.7237	
26	1.64	44.9	50.0802	-5.1802	
27	1.85	44.1	46.8147	-2.7147	
28	1.19	65.7	57.0778	8.6222	
29	1.32	59.1	55.0563	4.0437	

There are many large residuals. Those marked by an * appear to be unusually large.

11.47

x	y	x-rank	y-rank	(x-rank)(y-rank)
16.1	4.41	1	1	1
31.5	6.81	9	7	63
21.5	5.26	3	2	6
22.4	5.99	4	5	20
20.5	5.92	2	4	8
28.4	6.14	6	6	36
30.3	6.84	8	8	72
25.6	5.87	5	3	15
32.7	7.03	10	10	100
29.2	6.89	7	9	56
34.7	7.87	11	11	121
		66	66	497

$$r_s = \frac{497 - \frac{(66)(66)}{11}}{\sqrt{506 - \frac{(66)^2}{11}}\sqrt{506 - \frac{(66)^2}{11}}} = \frac{101}{\sqrt{110}\ \sqrt{110}} = .9182$$

11.49 a) For an x value two standard deviations above its mean, the standardized score for the predicted y would be (-.75)(2) = -1.5. Thus, the predicted y would be 1.5 standard deviations below the mean y. For an x value one standard deviation below its mean, the standardized score for the predicted y would be (-.75)(-1) = .75. Thus, the predicted y would be .75 standard deviations above the mean y.

b) When age six height is 50, $\hat{y} = 70 + .8\left(\frac{2.5}{1.7}\right)(50 - 46) = 74.71$

Chapter 12
Regression and Correlation: Inferential Methods

Section 1

12.1 a) $y = -5.0 + .017x$

b) When $x = 1000$, $y = -5 + .017(1000) = 12$
When $x = 2000$, $y = -5 + .017(2000) = 29$

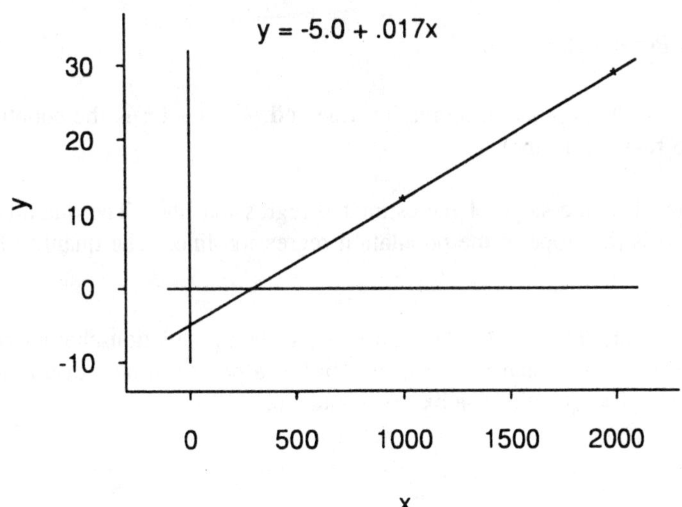

c) When $x = 2100$, $y = -5 + (.017)(2100) = 30.7$

d) .017

e) .017(100) = 1.7

f) When x = 500, y = −5 + .017(500) = 3.5
The model does not yield a usage value that is unreasonable (that is, negative). There is no information to suggest the model is adequate or inadequate for houses of this size.

12.3 a) The mean value of serum manganese when Mn intake is 4.0 is −2 + 1.4(4) = 3.6.

The mean value of serum manganese when Mn intake is 4.5 is −2 + 1.4(4.5) = 4.3.

b) $$\frac{5 - 3.6}{1.2} = 1.17$$

P(serum Mn over 5) = P(1.17 < z) = 1 − .8790 = .121

c) The mean value of serum manganese when MN intake is 5 is −2 + 1.4(5) = 5.

$$\frac{5 - 5}{1.2} = 0, \quad \frac{3.8 - 5}{1.2} = -1$$

P(serum Mn over 5) = P(0 < z) = .5

P(serum Mn below 3.8) = P(z < −1) = .1587

12.5 a) $y = \alpha + \beta x$ is the equation of the population regression line and $\hat{y} = a + bx$ is the equation of the least squares line (the estimated regression line).

b) The quantity b is a statistic. It is the slope of the estimated regression line. The quantity β is a population characteristic. It is the slope of the population regression line. The quantity b is an estimate of β.

c) $\alpha + \beta x^*$ is the true mean y-value for $x = x^*$. As such $\alpha + \beta x^*$ is a population characteristic. $a + bx^*$ is a point estimate of the mean y value when $x = x^*$ or $a + bx^*$ is a point estimate of an individual y value to be observed when $x = x^*$. The quantity $a + bx^*$ is a statistic.

12.7 a) $\hat{y} = 47.800050 - .16667x$

b) $\hat{y} = 47.800050 - .116667(150) = 47.800050 - 17.500050 = 30.3$

c) $\hat{y} = 47.800050 - .116667(120) = 47.800050 - 14.000040 = 33.80001$

d) Since 250 is well beyond the range of x values for which data has been observed, there would be danger in extrapolating the least-squares equation out to such a distant point.

12.9 a)

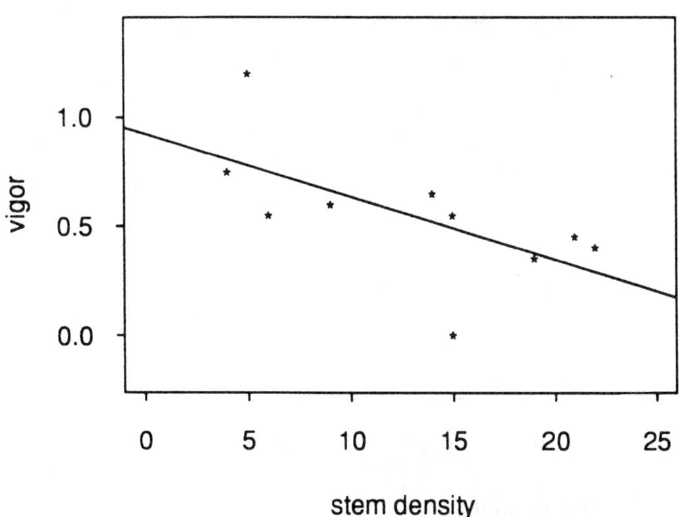

b)

$$b = \frac{59.95 - \left[\frac{(130)(5.5)}{10}\right]}{2090 - \left[\frac{(130)^2}{10}\right]} = \frac{-11.55}{400} = -0.02888$$

$a = .55 - (-.02888)(13) = 0.9254$

The equation of the estimated regression line is $\hat{y} = 0.9254 - .02888x$.

c) -0.02888

d) $0.9254 - .02888(17) = .4344$

12.11 a)

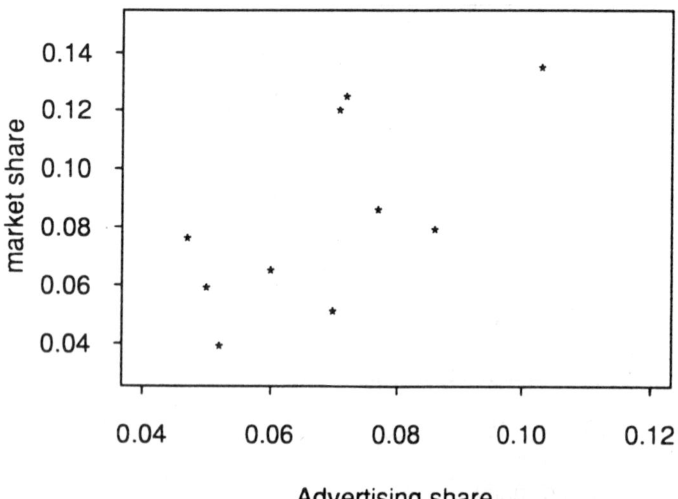

There seems to be a general tendency for y to increase at a constant rate as x increases. However, there is also quite a bit of variability in the y values. It is questionable whether a simple linear regression will adequately describe the relationship between x and y.

b) Summary values are: n = 10, $\Sigma x = 0.688$, $\Sigma x^2 = 0.050072$, $\Sigma y = .835$, $\Sigma y^2 = .079491$, $\Sigma xy = 0.060861$.

$$b = \frac{0.060861 - \left[\frac{(.688)(.835)}{10}\right]}{0.050072 - \left[\frac{(.688)^2}{10}\right]} = \frac{.003413}{.0027376} = 1.246712$$

$a = .0835 - (1.246712)(.0688) = -.002274$

The equation of the estimated regression line is $\hat{y} = -.002274 + 1.246712x$.

The predicted market share, when advertising share is .09, would be $-.002274 + 1.246712(.09) = .10993$.

c)

$$SSTo = .079491 - \left[\frac{(.835)^2}{10}\right] = .0097685$$

$SSResid = .079491 - (-.002274)(.835) - (1.246712)(.060861) = .0055135$

$$r^2 = 1 - \frac{.0055135}{.0097685} = 1 - .564 = .436$$

This means that 43.6% of the total variability in market share (y) can be explained by the simple linear regression model relating market share and advertising share (x).

d)

$$s_e = \sqrt{\frac{.0055135}{8}} = \sqrt{.000689} = .0263 \text{ with 8 d.f.}$$

Section 2

12.13 a) σ represents the standard deviation of the random deviation e. It is the typical deviation of the errors about the true regression line. σ_b represents the standard deviation of the statistic b. It is the typical deviation of b from its mean value β.

b) s_b is an estimate of σ_b based on the sample data. It is an estimate of the typical deviation of b from its mean value β.

12.15 For the 18 data values actually used,

$$\Sigma(x - \bar{x})^2 = 1158.33 - \frac{(143.1)^2}{18} = 1158.3 - 1137.645 = 20.685$$

For the 18 proposed data values, $\Sigma x = 135$, $\Sigma x^2 = 1042.5$ and

$$\Sigma(x - \bar{x})^2 = 1042.5 - \frac{(135)^2}{18} = 1042.5 - 1012.5 = 30.0$$

Since $\Sigma(x - \bar{x})^2$ is larger for the proposed data values and therefore making s_b smaller, it would have been better to use the proposed data values if the objective had been to estimate β as accurately as possible.

12.17 a) β denotes the average change in sale price associated with an average unit ($1,000) increase in appraised price.

H_0: $\beta = 1$ H_a: $\beta > 1$

Test statistic: $t = \dfrac{b - 1}{s_b}$ with d.f. = 12

For significance level .01, reject H_0 if $t > 2.68$.

$$t = \frac{(1.25 - 1)}{.30} = 0.83$$

Since .83 is not in the rejection region, the null hypothesis is not rejected. The data does not support the conclusion that the average change in sales price associated with a one unit increase in appraised price exceeds one.

b) Since the P–value of .0017 is less than the significance level of .01, there does appear to be a useful linear relationship between appraised value and sale price for this region.

c) Yes, the conclusion would have changed because the P–value would be greater than the level of significance.

12.19 a)

$$s_e = \sqrt{\frac{1235.47}{13}} = \sqrt{95.036} = 9.7486$$

$$s_b = \frac{9.7486}{\sqrt{4024.2}} = \frac{9.7486}{63.4366} = 0.1537$$

b) The 95% confidence interval for β is 2.5 ± (2.16)(.1537) = 2.5 ± .33 = (2.17, 2.83).

c) The interval is relatively narrow, and so, yes, the interval suggests that β has been precisely estimated.

12.21 a)

$$\Sigma xy - \left[\frac{(\Sigma x)(\Sigma y)}{n}\right] = 44194 - \left[\frac{(50)(16705)}{20}\right] = 2431.5$$

$$\Sigma x^2 - \left[\frac{(\Sigma x)^2}{n}\right] = 150 - \left[\frac{(50)^2}{20}\right] = 25$$

$$b = \frac{2431.5}{25} = 97.26, \quad a = 835.25 - (97.26)(2.5) = 592.1$$

b) $\hat{y} = 592.1 + 97.26(2) = 786.62$. The corresponding residual is $(y - \hat{y}) = 757 - 786.62 = -29.62$.

c) SSResid = 14194231 – 592.1(16705) – 97.26(44194) = 4892.06

$$s_e = \sqrt{\frac{4892.06}{18}} = \sqrt{271.781} = 16.4858$$

$$s_b = \frac{16.4858}{\sqrt{25}} = 3.2972$$

The 99% confidence interval for β, the true average change in oxygen usage associated with a one-minute increase in exercise time is 97.26 ± (2.88)(3.2972) = 97.26 ± 9.50 = (87.76, 106.76).

12.23 a) Let β denote the expected change in sales revenue associated with a one unit increase in advertising expenditure.

H_0: $\beta = 0$ $\quad H_a$: $\beta \neq 0$

Test statistic: $t = \dfrac{b}{s_b}$ with d.f. = 13

For significance level .05, reject H_0 if $t < -2.16$ or if $t > 2.16$.

$$t = \frac{52.57}{8.05} = 6.53$$

Since the calculated t of 6.53 exceeds the critical t of 2.16, H_0 is rejected. The simple linear regression model does provide useful information for predicting sales revenue from advertising expenditures.

b) H_0: $\beta = 40$ H_a: $\beta > 40$

Test statistic: $t = \dfrac{b - 40}{s_b}$ with d.f. = 13

For significance level .01, reject H_0 if $t > 2.65$.

$$t = \frac{(52.57 - 40)}{8.05} = 1.56$$

Since the calculated t of 1.56 does not exceed the critical t of 2.65, H_0 is not rejected. The data suggests that the average change in sales revenue associated with a one unit increase in advertising expenditure does not exceed 40 thousand dollars.

12.25 Let β denote the average change in milk pH associated with a one unit increase in temperature.

H_0: $\beta = 0$ H_a: $\beta < 0$

The test statistic is: $t = \dfrac{b}{s_b}$ with d.f. = 14.

At level of significance .01, reject H_0 if $t < -2.62$.

Computations: $n = 16$, $\Sigma x = 678$, $\Sigma y = 104.54$,

$$\Sigma xy - \left[\frac{(\Sigma x)(\Sigma y)}{n}\right] = 4376.36 - \left[\frac{(678)(104.54)}{16}\right] = -53.5225$$

$$\Sigma x^2 - \left[\frac{(\Sigma x)^2}{n}\right] = 36056 - \left[\frac{(678)^2}{16}\right] = 7325.75$$

$$b = \frac{-53.5225}{7325.75} = -.0073$$

$a = 6.53375 - (-.0073)(42.375) = 6.8431$

SSResid = $683.447 - 6.8431(104.54) - (-.0073)(4376.36) = .016754$

$$s_e = \sqrt{\frac{.016754}{14}} = \sqrt{.001197} = .0346$$

$$s_b = \frac{.0346}{\sqrt{7325.75}} = .000404$$

$$t = \frac{-.0073}{.000404} = -18.07$$

Since the calculated t of -18.07 is less than the critical t of -2.62, H_0 is rejected. There is sufficient evidence in the sample to conclude that there is a negative (inverse) linear relationship between temperature and pH.

12.27 From the MINITAB output, the value of b = 0.0068006 and s_b = 0.0002618.

a) Let β denote the expected change in extraction time associated with a one unit increase in pressure.

H_0: $\beta = 0.006$ H_a: $\beta \neq 0.006$

The test statistic is: $t = \dfrac{b - 0.006}{s_b}$ with d.f. = 8.

For level of significance .10, reject H_0 if $t < -1.86$ or if $t > 1.86$.

$$t = \frac{(.0068006 - .006)}{.0002618} = 3.058$$

Since the t calculated value of 3.058 falls into the rejection region, the null hypothesis is rejected. The data does contradict the prior belief of the investigators. It appears that $\beta \neq .006$.

b) From the table of t critical values, using degrees of freedom of 8, the P-value is found to be .02 > P-value > .01.

c) The 95% confidence interval is:

$.0068006 \pm (2.31)(.0002618) = .0068006 \pm .000605 = (.006196, .007405)$.

With 95% confidence, the expected change in extraction time associated with a one unit increase in pressure is estimated to be between .0062 minutes and .0074 minutes. The interval is quite narrow and so it does appear that β has been precisely estimated.

Section 3

12.29 If the request is for a confidence interval for β the wording would be like "estimate the *change* in the average y value associated with a one unit increase in the x variable." If the request is for a confidence interval for $\alpha + \beta x^*$ the wording would be like "estimate the *average* y value when the value of the x variable is x*."

12.31 a) $1.0357 \pm (3.06)(.014) \Rightarrow 1.0357 \pm .0428 \Rightarrow (.9929\ ,\ 1.0785)$

b) $1.0357 \pm (3.06)\sqrt{(.0420)^2 + (.0140)^2} = 1.0357 \pm (3.06)(.0443) = 1.0357 \pm .1356 = (.9001\ ,\ 1.1713)$

12.33 a) When x* = 40, y = a+bx* = 6.843345 − .00730608(40) = 6.5511.

$$s_{a+b(40)} = .0356\sqrt{\frac{1}{16} + \frac{(40 - 42.375)^2}{7325.75}} = .008955$$

The 95% confidence interval for $\alpha + \beta(40)$ is

$6.5511 \pm (2.15)(.008955) = 6.5511 \pm .0193 = (6.5318, 6.5704)$.

b) When x* = 35, a+bx* = 6.843345 − .00730608(35) = 6.5876.

$$s_{a+b(35)} = .0356\sqrt{\frac{1}{16} + \frac{(35 - 42.375)^2}{7325.75}} = .009414$$

The 99% confidence interval for $\alpha + \beta(35)$ is

$6.5876 \pm (2.98)(.009414) = 6.5876 \pm .0281 = (6.5595, 6.6157)$.

c) The value 90 is outside the range of x values for which data was collected and is quite far from the mean $\bar{x} = 42.375$. Hence the confidence interval for $\alpha + \beta(90)$ would be quite wide. Thus, I would not recommend constructing the requested interval.

12.35 The midpoint of the prediction interval should be $a + b\bar{x}$. However, from the midpoint of the interval given in (145 + 572)/2 = 358.5, not 277. So yes, there is an inconsistency in the information given.

12.37 Since 17 is farther away from $\bar{x} = 19.21$ than is 20, the confidence interval with x* = 17 would be wider. The same would be true for a prediction interval.

12.39 a) From 12.32 (a), the 95% prediction interval for time when pressure is 200 is (3.2416, 3.7624). The 95% prediction interval for time when pressure is 250 is

$3.842 \pm 2.31(.1116) = 3.842 \pm .258 = (3.584, 4.100)$.

Therefore, the pair of intervals form a set of simultaneous prediction intervals with prediction level of at least 90%.

b) The simultaneous prediction level would be at least [100 − 3(1)]% = 97%.

Section 4

12.41 The quantity r is a statistic as its value is calculated from the sample. It is a measure of how strongly the sample x and y values are related. The value of r is an estimate of ρ. The quantity ρ is a population characteristic. It measures the strength of association between the x and y values in the population.

12.43 a) Let ρ denote the correlation coefficient between time spent watching television and grade point average in the population from which the observations were selected.

H_0: $\rho = 0$ H_a: $\rho < 0$

Test statistic:

$$t = \frac{r}{\sqrt{\frac{(1 - r^2)}{(n - 2)}}}$$ with degree of freedom equal to 526.

For a significance level of .01 and df = 526, the t critical value is approximately −2.33.
Reject H_0 if $t < -2.33$.

$$t = \frac{-.26}{\sqrt{\frac{[1 - (-.26)^2]}{526}}} = -6.176$$

Since −6.176 falls in the rejection region, the null hypothesis is rejected. The data does support the conclusion that there is a negative correlation in the population between the two variables, time spent watching television and grade point average.

b) The coefficient of determination measures the proportion of observed variation in grade point average explained by the regression on time spent watching television. This value would be $(-.26)^2 = .0676$. Thus only 6.76% of the observed variation in grade point average would be explained by the regression. This is not a substantial percentage.

12.45 a) Let ρ denote the true correlation between particulate pollution and luminance.

H_0: $\rho = 0$ H_a: $\rho > 0$

$$t = \frac{r}{\sqrt{\frac{(1 - r^2)}{(n - 2)}}}$$ with d.f. = 13

Since no significance level was specified, .01 will be used. Reject H_0 if $t > 2.65$.

$$\Sigma xy - \left[\frac{(\Sigma x)(\Sigma y)}{n}\right] = 22265 - \left[\frac{(860)(348)}{15}\right] = 2313$$

$$\Sigma x^2 - \left[\frac{(\Sigma x)^2}{n}\right] = 56700 - \left[\frac{(860)^2}{15}\right] = 7393.333$$

$$\Sigma y^2 - \left[\frac{(\Sigma y)^2}{n}\right] = 8954 - \left[\frac{(348)^2}{15}\right] = 880.4$$

$$r = \frac{2313}{\sqrt{(7393.333)(880.4)}} = .9066$$

$$t = \frac{.9066}{\sqrt{\frac{[1 - (.9066^2)]}{13}}} = 7.75$$

Since the calculated t of 7.75 exceeds the critical t of 2.65, H_0 is rejected. The data supports the conclusion that there is a positive correlation between particulate pollution and luminance.

b) $r^2 = (.9066)^2 = .822$

12.47 From the data, the following are computed:

$n = 6$, $\Sigma x = 111.71$, $\Sigma x^2 = 2724.7943$, $\Sigma y = 2.9$, $\Sigma y^2 = 1.6572$, $\Sigma xy = 63.915$,

$$\Sigma xy - \left[\frac{(\Sigma x)(\Sigma y)}{n}\right] = 63.915 - \left[\frac{(111.71)(2.9)}{6}\right] = 9.9218$$

$$\Sigma x^2 - \left[\frac{(\Sigma x)^2}{n}\right] = 2724.7943 - \left[\frac{(111.71)^2}{6}\right] = 644.9403$$

$$\Sigma y^2 - \left[\frac{(\Sigma y)^2}{n}\right] = 1.6572 - \left[\frac{(2.9)^2}{6}\right] = .2555$$

Let ρ denote the correlation between stiffness and thickness.

H_0: $\rho = 0$ H_a: $\rho \neq 0$

$$t = \frac{r}{\sqrt{\frac{(1 - r^2)}{(n - 2)}}} \quad \text{with d.f.} = 4$$

For significance level 0.05, reject H_0 if $t < -2.78$ or if $t > 2.78$.

$$r = \frac{9.9218}{\sqrt{(644.9403)(.2555)}} = .773$$

$$t = \frac{.773}{\sqrt{\frac{[1 - (.773)^2]}{4}}} = 2.44$$

Since the t calculated value of 2.44 does not fall into the rejection region, the null hypothesis is not rejected. The data does not support the conclusion that there is a linear relationship between stiffness and thickness. The r value of .773 certainly "looks good." However, when n is quite small, an r value of large magnitude can easily result even when $\rho = 0$.

12.49 a) With level of significance .001 and P-value of .00032, the null hypothesis would be rejected. It would be concluded that ρ is not zero.

b) The conclusion in (a) is that $\rho \neq 0$. The small P-value does not indicate what value ρ may have, other than it is not zero. It does not indicate that ρ is large.

Section 5

12.51 a) With the exception of one, all have values from 4.2 to 4.9.

b) The ten standardized residuals are

$$\frac{4.72}{2.138} = 2.208 \qquad \frac{-.94}{4.358} = -2.16$$

$$\frac{3.68}{4.232} = 0.870 \qquad \frac{6.66}{4.358} = 1.528$$

$$\frac{-2.85}{4.440} = -.642 \qquad \frac{-1.33}{4.956} = -.268$$

$$\frac{-.02}{4.421} = -.005 \qquad \frac{-9.41}{4.203} = -2.239$$

$$\frac{-.44}{4.433} = -.099 \qquad \frac{-.11}{4.429} = -.025$$

c)

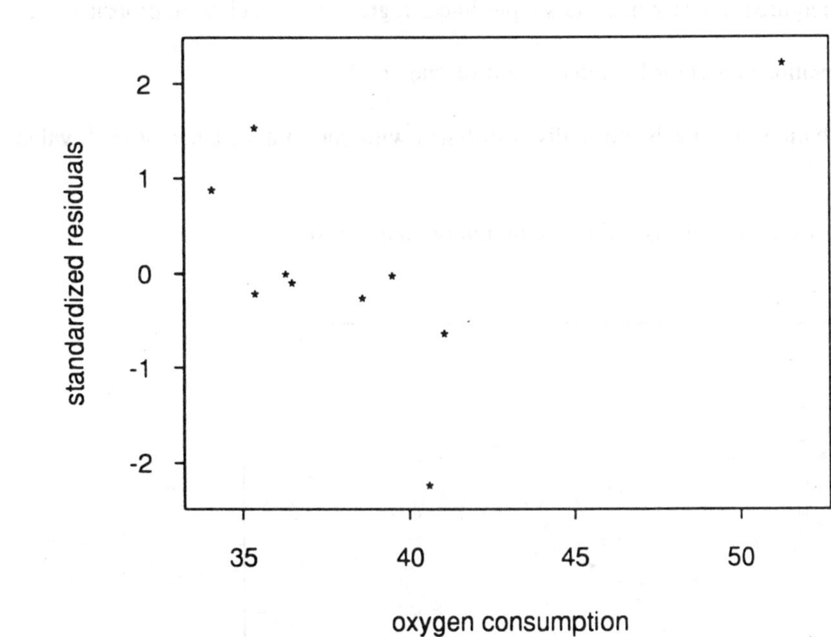

There is one point that is quite distant from the remaining points. It has a large residual as well as a value for oxygen that is much larger than the other oxygen readings. The remaining residuals appear to have a linear relationship with a negative slope. One suspects that the one discrepant data point might be having an undue influence on the estimated regression line.

d)

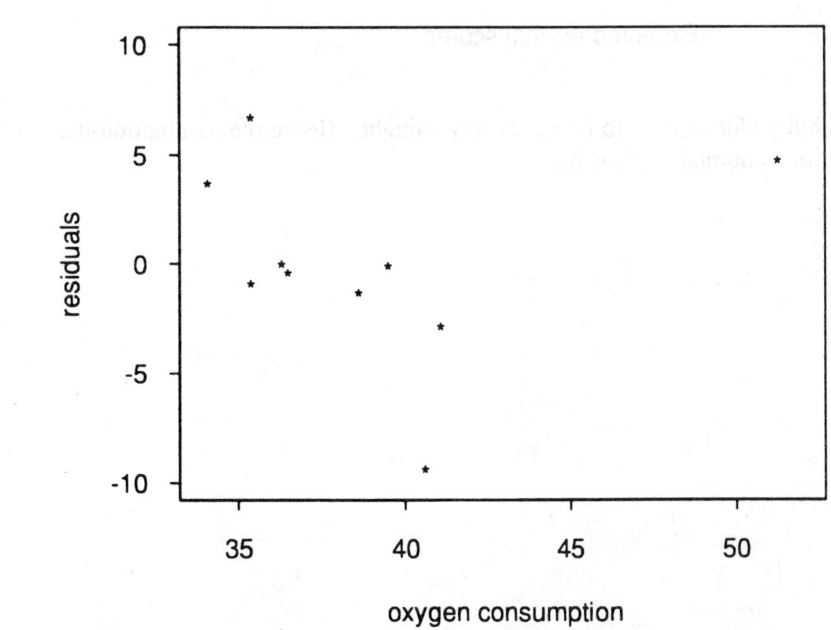

The residual plot exhibits the same basic features as the standardized residual plot.

12.53 a) The assumptions required in order that the simple linear regression model be appropriate are:

(i) the observations on vigor be independent of one another.

(ii) the distribution of vigor be normally distributed with constant variance at each value of stem density.

(iii) the mean value of vigor is a linear function of stem density.

b)

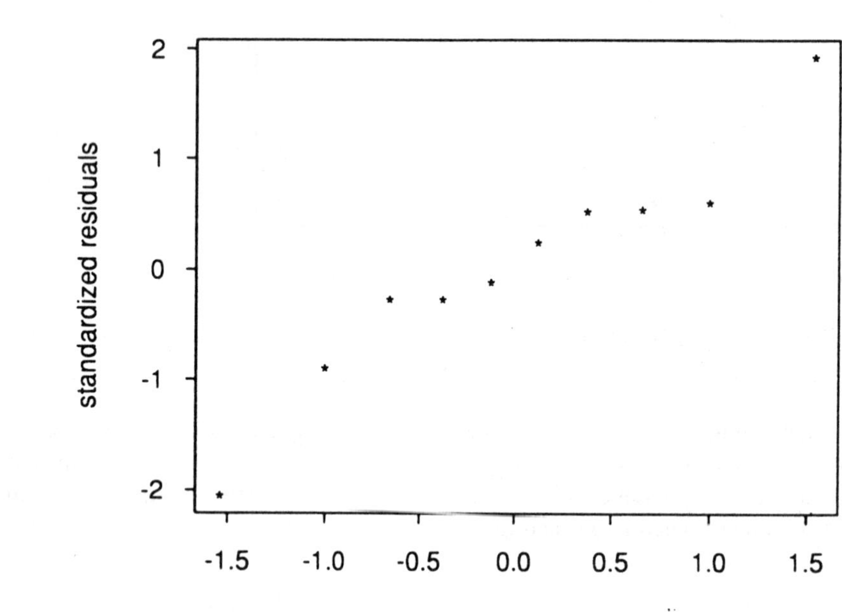

The normal probability plot appears to be reasonably straight. Hence the assumption that the random deviation distribution is normal is plausible.

c)

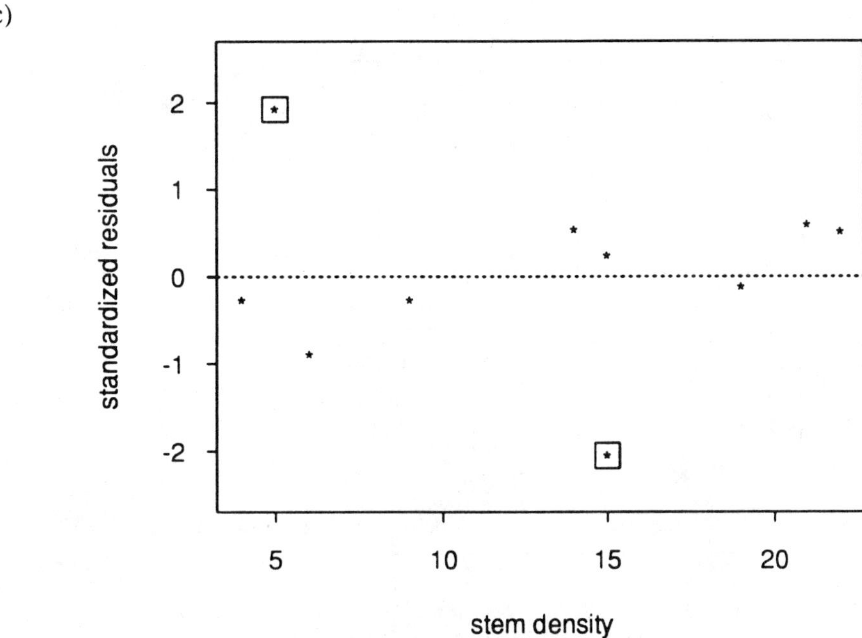

There are two residuals that are large relative to the other residuals. The corresponding points are enclosed in boxes on the graph above.

d) Except for the two large residuals, the negative residuals are associated mainly with small x values, and the positive residuals are associated with large x values. This would cause one to question the appropriateness of a simple linear regression model.

12.55

Year	X	Y	Y-Pred	Residual
1963	188.5	2.26	1.750	0.51000
1964	191.3	2.60	2.478	0.12200
1965	193.8	2.78	3.128	−0.34800
1966	195.9	3.24	3.674	−0.43400
1967	197.9	3.80	4.194	−0.39400
1968	199.9	4.47	4.714	−0.24400
1969	201.9	4.99	5.234	−0.24400
1970	203.2	5.57	5.572	−0.00200
1971	206.3	6.00	6.378	−0.37800
1972	208.2	5.89	6.872	−0.98200
1973	209.9	8.64	7.314	1.32600

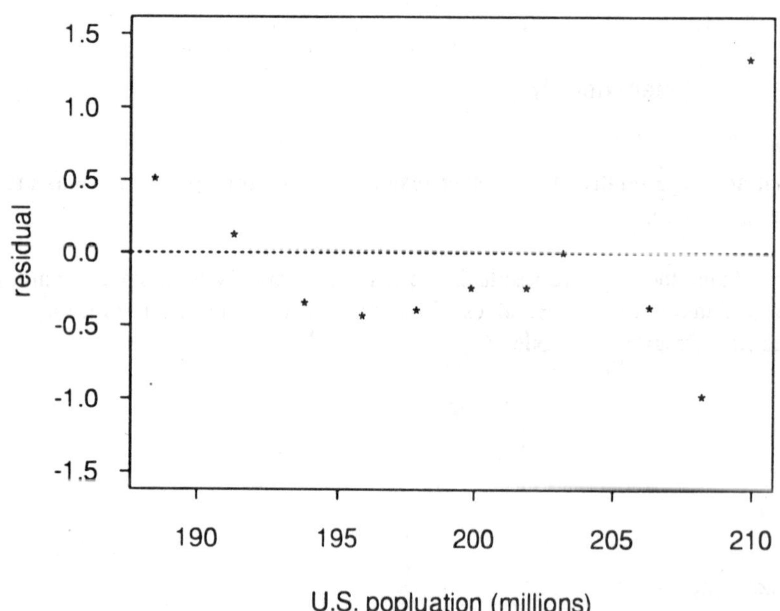

The residuals are positive in 1963 and 1964, then they are negative from 1965 through 1972, followed by a positive residual in 1973. The residuals exhibit a pattern in the plot and thus the plot casts doubt on the appropriateness of the simple linear regression model.

Section 6

12.57 a) $415.11 - 6.6(20) - 4.5(40) = 103.11$

b) $415.11 - 6.6(18.9) - 4.5(43) = 96.87$

12.59 a) $.69 + .47(2) + .00041(.1) - .72(1.2) + .023(6) = 9.634$

b) It would increase, since the value of β_2 is positive.

12.61 a) The value .469 is an estimate of the expected change in the mean score of students associated with a one unit increase in the student's expected score holding time spent studying and student's grade point average constant.

b) H_0: $\beta_1 = \beta_2 = \beta_3 = 0$

H_a: At least one of the three β_i's is not zero.

Test statistic: $F = \dfrac{R^2 / k}{(1 - R^2) / [n - (k + 1)]}$

With numerator d.f. = 3 and denominator d.f. = 103, Appendix Table IV(a) gives the level .05 critical value as approximately 2.70. Hence, reject H_0 if $F > 2.70$.

$$F = \frac{.686/3}{.314/103} = 75.01$$

Since 75.01 > 2.70, the null hypothesis is rejected. The data suggests that there is a useful linear relationship between exam score and at least one of the three predictor variables.

c) The 95% confidence interval for β_2 is $3.369 \pm (1.98)(.456) = 3.369 \pm .903 = (2.466, 4.272)$.

d) The point prediction would be $2.178 + .469(75) + 3.369(8) + 3.054(2.8) = 72.856$.

e) The prediction interval would be $72.856 \pm (1.98)\sqrt{s_e^2 + (1.2)^2}$.

To determine s_e^2, proceed as follows. From the definition of R^2, it follows that SSResid = $(1 - R^2)$SSTo.

So SSResid = (1 − .686)(10,200) = 3202.8. Then

$$s_e^2 = \frac{3202.8}{103} = 31.095.$$

The prediction interval becomes

$$72.856 \pm (1.98)\sqrt{31.095 + (1.2)^2} = 72.856 \pm (1.98)(5.704) = 72.856 \pm 11.294 = (61.562, 84.150).$$

12.63 a) H_0: $\beta_1 = \beta_2 = \ldots = \beta_8 = 0$

H_a: At least one of the eight β_i's is not zero.

Test statistic: $F = \dfrac{R^2 / k}{(1 - R^2) / [\, n - (k + 1)]}$

With numerator d.f. = 8 and denominator d.f. = 55, Appendix Table IV(a) gives the level .01 critical value as approximately 2.85. Hence, reject H_0 if $F > 2.85$.

$$F = \frac{.467/8}{.533/5} = 6.024$$

Since the calculated F of 6.024 exceeds the critical F value of 2.85, the null hypothesis is rejected. The model containing the eight variables appears to be a useful one. That is, at least one of the eight variables is useful for predicting labor efficiency index.

b) The P-value would be less than .001.

12.65 a) Estimated mean value of y = $.067 + .054x - .00052x^2$

b) For x = 20, estimated mean value of y is $.067 + .054(20) - .00052(20)^2 = .939$.

For x = 40, estimated mean value of y is $.067 + .054(40) - .00052(40)^2 = 1.395$.

c)

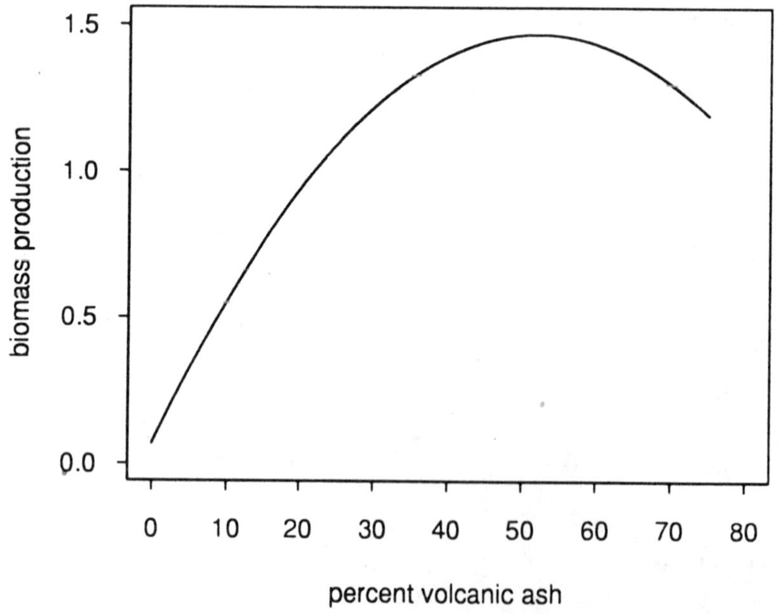

biomass production

1.5
1.0
0.5
0.0

0 10 20 30 40 50 60 70 80

percent volcanic ash

The point at which increasing percent volcanic ash begins to have a detrimental effect on average biomass is about 52.

12.67 a) The 95% confidence interval for β_3 is $-9.378 \pm (2.08)(4.356) = -9.378 \pm 9.06 = (-18.438, -.318)$.

With 95% confidence, the expected change in number of fish associated with a one unit change in sea state, while holding the other variable constant, is estimated to be between −18.438 and −.318.

b) $-2.179 \pm (1.72)(1.087) = -2.179 \pm 1.87 = (-4.049, -.309)$

Supplementary Exercises

12.69 a) $b = -.640$

b) $\hat{y} = 106.3 - .64(40) = 106.3 - 25.6 = 80.7$

c) Because the estimated slope is negative, the value of r is the negative square root of (.47), which is −.6856.

Let ρ denote the correlation coefficient between green biomass concentration and elapsed time since snowmelt.

$H_0\colon \rho = 0 \quad H_a\colon \rho \neq 0$

Test statistic:

$$t = \frac{r}{\sqrt{\frac{(1 - r^2)}{(n - 2)}}} \quad \text{with d.f.} = 56$$

$$t = \frac{-.6856}{\sqrt{(1 - .47) / 56}} = \frac{-.6856}{.0973} = -7.05$$

The P-value equals the area under a t curve (with 56 d.f.) to the left of −7.05. From the table of t critical values, this area (P-value) is less than .001. Thus, the data does suggest that there is a useful linear relationship between elapsed time since snowmelt and green biomass concentration.

12.71 a) When $x^* = 20$, $a+b(20) = 2.2255 + .1521(20) = 5.2675$.

$$s^2_{a+b(20)} = .1187\left[\frac{1}{11} + \frac{(20 - 26.627)^2}{342.622}\right] = .026$$

$$s_{a+b(20)} = .1613$$

The 95% confidence interval for the mean tensile modulus when bound rubber content is 20% is:

$5.2675 \pm 2.26(.1613) = 5.2675 \pm .3645 = (4.903, 5.632)$.

b) From (a), y = 5.2675 and the 95% prediction interval is

$$5.2675 \pm 2.26\sqrt{.1187 + (.1613)^2} = 5.2675 \pm .8597 = (4.4078, 6.1272)$$

c) The request is to estimate the "true *mean* tensile strength" at a given value of x, not to estimate the "*change* in mean tensile strength" associated with a one-unit increase in x.

12.73 From the data: n = 9, $\Sigma xy - \left[\frac{(\Sigma x)(\Sigma y)}{n}\right] = 1390.2667$,

$$\Sigma x^2 - \left[\frac{(\Sigma x)^2}{n}\right] = 93.429,\ \Sigma y^2 - \left[\frac{(\Sigma y)^2}{n}\right] = 295534$$

$$r = \frac{1390.2667}{\sqrt{(93.429)(295534)}} = .2646.$$

Let ρ denote the correlation between eye weight and cornea thickness.

H_0: ρ = 0 H_a: ρ > 0

$$t = \frac{r}{\sqrt{\frac{(1 - r^2)}{(n - 2)}}} \quad \text{with d.f.} = 7$$

For significance level 0.05, reject H_0 if t > 1.90.

$$t = \frac{.2646}{\sqrt{\frac{[1 - (.2646)^2]}{7}}} = .73$$

Since the t calculated value of .73 does not fall in the rejection region, the null hypothesis is not rejected. The data does not support the conclusion of a positive correlation between eye weight and cornea thickness.

12.75 a)

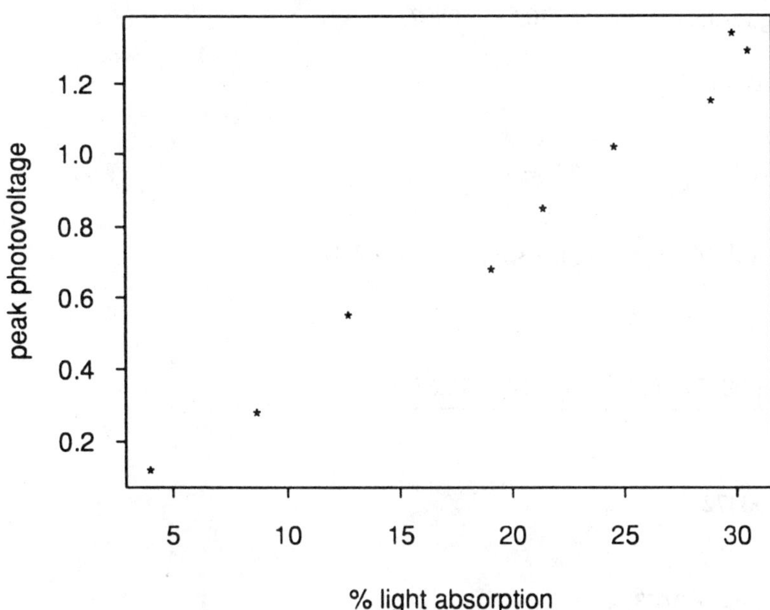

The plot suggests that a straight line model might adequately describe the relationship between percent light absorption and peak photovoltage.

b)

$$\Sigma xy - \left[\frac{(\Sigma x)(\Sigma y)}{n}\right] = 178.683 - \left[\frac{(179.7)(7.28)}{9}\right] = 33.326$$

$$\Sigma x^2 - \left[\frac{(\Sigma x)^2}{n}\right] = 4334.41 - \left[\frac{(179.7)^2}{9}\right] = 746.4$$

$$\Sigma y^2 - \left[\frac{(\Sigma y)^2}{n}\right] = 7.4028 - \left[\frac{(7.28)^2}{9}\right] = 1.514$$

$$b = \frac{33.326}{746.4} = .044649$$

$$a = .809 - .044649(19.667) = -0.08259$$

$$\hat{y} = -0.08259 + .044649x$$

c) $$r^2 = \frac{(33.326)^2}{(746.4)(1.514)} = .983$$

d) When $x^* = 19.1$, $y = -.08259 + .044649(19.1) = .7702$.
The corresponding residual is $.68 - .7702 = -.0902$.

e) $H_0: \beta = 0 \quad H_a: \beta \neq 0$

The test statistic is: $t = \dfrac{b}{s_b}$ with d.f. = 7

For significance level .05, reject H_0 if $t < -2.37$ or $t > 2.37$.

$b = .044649$

$$s_e^2 = \frac{7.4028 - (-.08259)(7.28) - .044649(178.683)}{7}$$

$$s_e^2 = \frac{.02604}{7} = .00372$$

$$s_b^2 = \frac{.00372}{746.4} = 4.984 \bullet 10^{-6},\ s_b = .00223$$

$$t = \frac{.044649}{.00223} = 20.00$$

Since the calculated t of 20 exceeds the critical t of 2.37, H_0 is rejected. The data does support the conclusion that there is a useful linear relationship between percent light absorption and peak photovoltage.

f) In the absence of a specified confidence level, 95% will be used. The 95% confidence interval for the average change in peak photovoltage associated with a 1% increase in light absorption is $.044649 \pm 2.37(.00223) = .044649 \pm .00529 = (.039359, .049939)$

g) In the absence of a specified confidence level, 95% will be used.
When $x^* = 20$, $\hat{y} = -.08259 + .044649(20) = .8104$

$$s_{a+b(20)} = .061\sqrt{\frac{1}{9} + \frac{(20 - 19.667)^2}{746.4}} = .0204$$

The 95% confidence interval of true average peak photovoltage when percent light absorption is 20 is $.8104 \pm 2.37(.0204) = .8104 \pm .0483 = (.7621, .8587)$.

12.77 a) The summary values are: $n = 10$, $\Sigma x = 25$, $\Sigma x^2 = 145$, $\Sigma y = -.4$, $\Sigma y^2 = 43.88$, $\Sigma xy = 55.5$.

$$b = \frac{55.5 - \left[\frac{(25)(-.4)}{10}\right]}{145 - \left[\frac{(25)^2}{10}\right]} = \frac{56.5}{82.5} = .68485$$

$a = -.04 - .68485(2.5) = -1.7521$

The equation of the estimated regression line is $\hat{y} = -1.7521 + .68485x$.

b) SSResid = 43.88 − (−1.7521)(−.4) − (.68485)(55.5) = 5.169985

$$s_e^2 = \frac{5.169985}{8} = .646248$$

$$s_a^2 = s_{a+b(0)}^2 = s_e^2\left[\frac{1}{n} + \frac{\bar{x}^2}{\Sigma x^2 - \left[\frac{(\Sigma x)^2}{n}\right]}\right] = .646248\left[\frac{1}{10} + \frac{(2.5)^2}{82.5}\right]$$

$s_a^2 = .113583$, $s_a = .337$

H_0: $\alpha = 0$ H_a: $\alpha \neq 0$

The test statistic is: $t = \frac{a}{s_a}$ with d.f. = 8.

For significance level .05, reject H_0 it $t < -2.31$ or if $t > 2.31$.

$$t = \frac{-1.7521}{.337} = -5.20$$

Since the t calculated value of −5.20 falls in the rejection region, the null hypothesis is rejected. The data suggests that the y intercept of the true regression line differs from zero.

c) The 95% confidence interval for α is $-1.7521 \pm (2.31)(.337) = -1.7521 \pm .7785 = (-2.5306, -.9736)$

Since the interval does not contain the value zero, zero is not one of the plausible values for α.

12.79 Summary values for Leptodactylus ocellatus: $n = 9$, $\Sigma x = 64.2$, $\Sigma x^2 = 500.78$, $\Sigma y = 19.6$, $\Sigma y^2 = 47.28$, $\Sigma xy = 153.36$.

From these: $b = .31636$, SSResid $= .3099$, $\Sigma(x - \bar{x})^2 = 42.82$

Summary values for Bufa marinus: $n = 8$, $\Sigma x = 55.9$, $\Sigma x^2 = 425.15$, $\Sigma y = 21.6$, $\Sigma y^2 = 62.92$, $\Sigma xy = 163.63$

From these: $b = .35978$, SSResid $= .1279$, $\Sigma(x - \bar{x})^2 = 34.549$

$$s^2 = \frac{.3099 + .1279}{9 + 8 - 4} = \frac{.4378}{13} = .0337$$

H_0: $\beta = \beta'$ $\quad H_a$: $\beta \neq \beta'$

The test statistic is:

$$t = \frac{b - b'}{\sqrt{\dfrac{s^2}{SS_x} + \dfrac{s^2}{SS_x'}}} \quad \text{with d.f.} = 9 + 8 - 4 = 13.$$

For significance level .05, reject H_0 if $t < -2.16$ or if $t > 2.16$.

$$t = \frac{.31636 - .35978}{\sqrt{\dfrac{.0337}{42.82} + \dfrac{.0337}{34.549}}} = \frac{-.04342}{.04198} = -1.03$$

Since the t calculated value of -1.03 does not fall in the rejection region, the null hypothesis of equal regression slopes cannot be rejected. The data suggests that the slopes of the true regression lines for the two different frog populations may be identical.

12.81 Summary values are: $n = 8$,

$$\Sigma x^2 - \left[\frac{(\Sigma x)^2}{n}\right] = 42, \quad \Sigma y^2 - \left[\frac{(\Sigma y)^2}{n}\right] = 586.875, \quad \Sigma xy - \left[\frac{(\Sigma x)(\Sigma y)}{n}\right] = 1.5.$$

a) $b = \dfrac{1.5}{42} = .0357$

$a = 58.125 - (.0357)(4.5) = 57.964$

The equation of the estimated regression line is $\hat{y} = 57.964 + .0357x$.

Form the SAS output the values of a and b are read to be 57.964286 and 0.035714. So the equation of the estimated regression line using the SAS output would be $\hat{y}$ = 57.964286 + .035714x.

b) Let β denote the expected change in glucose concentration associated with a one day increase in fermentation time.

H_0: β = 0 H_a: β ≠ 0

The test statistic: $t = \dfrac{b}{s_b}$ with d.f. = 6.

For significance level 0.10, reject H_0 if t < −1.94 or if t > 1.94.

From the SAS output, s_b = 1.52599350 and t = .023.

Since the t calculated value of .023 does not fall in the rejection region, the null hypothesis is not rejected. The data does not indicate a linear relationship between fermentation time and glucose concentration.

c)

x	y	Pred-y	Residual
1	74	58.00	16.00
2	54	58.04	−4.04
3	52	58.07	−6.07
4	51	58.11	−7.11
5	52	58.14	−6.14
6	53	58.18	−5.18
7	58	58.21	−0.21
8	71	58.25	12.75

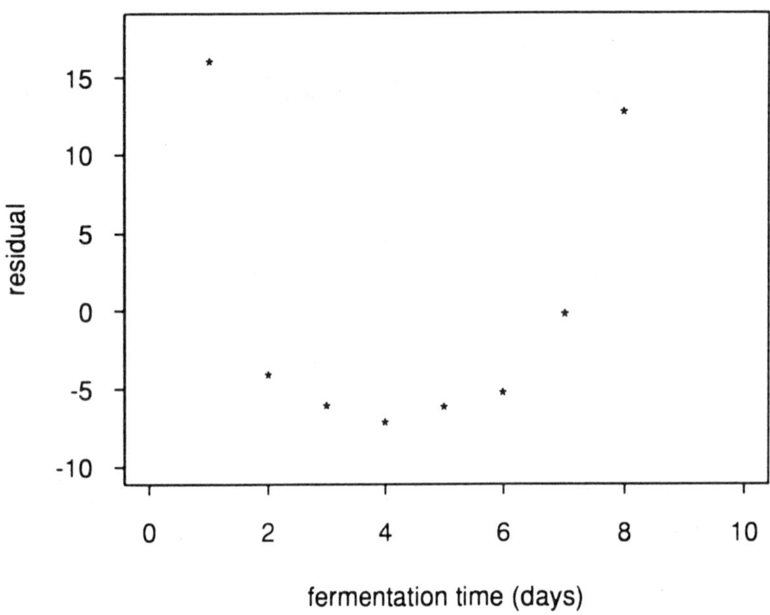

d) The residual plot has a very distinct curvilinear pattern which indicates that a simple linear regression model is not appropriate for describing the relationship between y and x. Instead, a model incorporating curvature should be fit.

12.83 a) The e_i's are the deviations of the observations from the population regression line, whereas the residuals are the deviations of the observations from the estimated regression line.

b) The simple linear regression model states that $y = \alpha + \beta x + e$. Without the random deviation e, the equation implies a deterministic model, whereas the simple linear regression model is probabilistic.

c) The quantity b is a statistic. Its value is known once the sample has been collected, and different samples result in different b values. Therefore, it does not make sense to test hypotheses about b. Only hypotheses about a population characteristic can be tested.

d) If $r = +1$ or -1, then each point falls exactly on the regression line and SSResid would equal zero. A true statement is that SSResid is always greater than or equal to zero.

e) The sum of the residuals must equal zero. Thus, if they are not all exactly zero, at least one must be positive and at least one must be negative. They cannot all be positive. Since there are some positive and no negative values among the reported residuals, the student must have made an error.

f) $SSTo = \Sigma(y - \bar{y})^2$ must be greater than or equal to $SSResid = \Sigma(y - \hat{y})^2$. Thus, the values given must be incorrect.

Chapter 13
The Analysis of Variance

Section 1

13.1 Each of the k population or treatment response distributions is normal and their variances are equal.

13.3 a) Let μ_1, μ_2, μ_3 and μ_4 denote the true average length of stay in a hospital for health plans 1, 2, 3 and 4 respectively.

H_0: $\mu_1 = \mu_2 = \mu_3 = \mu_4$

H_a: At least two of the four μ_i's are different.

b) The numerator d.f. = 4 − 1 = 3 and the denominator d.f. = 32 − 4 = 28. At $\alpha = 0.01$, reject H_0 if $F > 4.57$. Since the computed F of 4.37 does not exceed 4.57, H_0 is not rejected. Hence, it would be concluded that the average length of stay in the hospital is the same for the four health plans.

c) The numerator d.f. = 4 − 1 = 3 and the denominator d.f. = 32 − 4 = 28. Thus the critical F value remains the same as in (b). Therefore, the conclusion would be the same.

13.5 a) Let μ_i denote the mean satisfaction level for technique i (i = 1, 2, 3).

H_0: $\mu_1 = \mu_2 = \mu_3$

H_a: At least two of the three μ_i's are different.

Test statistic: $F = \dfrac{\text{MSTr}}{\text{MSE}}$

Rejection region: Numerator d.f. = k − 1 = 2 and denominator d.f. = N − k = 30. For α = .05, Appendix Table IV(a) gives 3.32 as the critical value. The null hypothesis will be rejected if F > 3.32.

$$F = \frac{MSTr}{MSE} = 4.12$$

Since 4.12 > 3.32, the null hypothesis is rejected. The data suggests that true mean satisfaction level depends on which interview technique is being used.

b) The denominator d.f. = 45 − 3 = 42. The critical value decreases as the denominator d.f. increases. Since the computed F of 4.12 exceeds the level .05 critical value when the denominator d.f. = 30, it will exceed the level .05 critical value when the denominator d.f. = 42. Hence, the null hypothesis would be rejected and the same conclusion as in (a) holds.

13.7 Let μ_i denote the mean tensile strength of wire of type i (i = 1, 2 ,..., 5).

H_0: $\mu_1 = \mu_2 = \mu_3 = \mu_4 = \mu_5$

H_a: At least two of the five μ_i's are different.

Test statistic: $F = \frac{MSTr}{MSE}$

Rejection region: Numerator d.f. = k − 1 = 4 and denominator d.f. = N − k = 15. For α = .01, Appendix Table IV(b) gives 4.89 as the critical value. The null hypothesis will be rejected if F > 4.89.

$$F = \frac{MSTr}{MSE} = \frac{2573.3}{1394.2} = 1.85$$

Since 1.85 < 4.89, the null hypothesis is not rejected. The data suggests that the mean tensile strength is the same for the five wire types.

13.9 a) numerator d.f. = 3 − 1 = 2
denominator d.f. = 24 + 24 + 20 − 3 = 65

b) Let μ_i denote the mean level of job satisfaction of employees in group i (i = 1, 2 ,3).

H_0: $\mu_1 = \mu_2 = \mu_3$

H_a: At least two of the three μ_i's are different.

Test statistic: $F = \frac{MSTr}{MSE}$

Rejection region: Numerator d.f. = k − 1 = 2 and denominator d.f. = N − k = 65. For α = .05, Appendix Table IV(a) gives 3.15 as the approximate critical value. The null hypothesis will be rejected if F > 3.15.

$$F = \frac{\text{MSTr}}{\text{MSE}} = 6.62$$

Since 6.62 > 3.15, the null hypothesis is rejected. The data suggests that the true mean satisfaction level depends on to which group an employee belongs.

c) $$\bar{\bar{x}} = \frac{[24(6.60) + 24(5.37) + 20(5.20)]}{(24 + 24 + 20)}$$

$$= \frac{(158.4 + 128.88 + 104)}{68} = \frac{391.28}{68} = 5.754$$

$\text{MSTr} = [24(6.6 - 5.754)^2 + 24(5.37 - 5.754)^2 + 20(5.2 - 5.754)^2]/2 = 26.854/2 = 13.427$

Since F = MSTr/MSE, it follows that MSE = MSTr/F. Since F = 6.62 and MSTr = 13.427, then MSE = 13.427/6.62 = 2.028.

13.11 Let μ_1, μ_2, μ_3, and μ_4 denote the true average phosphorus content of maize stored 0, 1, 2, and 4 months respectively.

H_o: $\mu_1 = \mu_2 = \mu_3 = \mu_4$

H_a : at least two among the four μ's are different.

Test statistic: $F = \frac{\text{MSTr}}{\text{MSE}}$

Rejection region: Numerator d.f. = k – 1 = 3 and denominator d.f. = N – k = 20. For α = .05, Appendix Table IV(a) gives 3.10 as the critical value. The null hypothesis will be rejected if F > 3.10.

$$\bar{\bar{x}} = \frac{362 + 350 + 350 + 349}{4} = \frac{1411}{4} = 352.75$$

$$\text{MSTr} = \frac{1}{3}[6(362 - 352.75)^2 + 6(350 - 352.75)^2 + 6(350 - 352.75)^2 + 6(349 - 352.75)^2]$$

$$= \frac{1}{3}[513.375 + 45.375 + 45.375 + 84.375] = \frac{688.50}{3} = 229.50$$

$$\text{MSE} = \frac{1}{20}[5(3)^2 + 5(4)^2 + 5(3)^2 + 5(2)^2] = \frac{190}{20} = 9.50$$

$$F = \frac{229.50}{9.50} = 24.158$$

Since 24.158 falls in the rejection region, the null hypothesis is rejected. The data supports the conclusion that the average phosphorus content of maize is not the same for each of the four storage times.

13.13 Let μ_1, μ_2, and μ_3 denote the true mean concentration determination for the three methods (i = 1, 2, 3).

$H_o : \mu_1 = \mu_2 = \mu_3$

H_a : at least two among the three μ's are different.

Test statistic: $F = \dfrac{MSTr}{MSE}$

Rejection region: Numerator d.f. = k − 1 = 2 and denominator d.f. = N − k = 6. For α = .05, Appendix Table IV(a) gives 5.14 as the critical value. The null hypothesis will be rejected if F > 5.14.

$$\bar{\bar{x}} = \frac{3(1.914) + 3(1.949) + 3(2.327)}{9} = \frac{18.57}{9} = 2.063$$

$$MSTr = \frac{1}{2}[3(1.914 - 2.063)^2 + 3(1.949 - 2.063)^2 + 3(2.327 - 2.063^2]$$

$$= \frac{.31468}{2} = .15734$$

$$MSE = \frac{1}{3}[(.212)^2 + (.095)^2 + (.198)^2] = \frac{.093173}{3} = 0.03106$$

$$F = \frac{.15734}{.03106} = 5.07$$

Since 5.07 < 5.14, the null hypothesis is not rejected. The data suggests that there is no difference in the mean concentration determination of the three methods.

13.15 Let μ_i denote the mean weight gain for salinity level i (i = 1, 2, 3).

H_o: $\mu_1 = \mu_2 = \mu_3$

H_a: At least two of the three μ_i's are different.

Test statistic: $F = \dfrac{MSTr}{MSE}$

Rejection region: Numerator d.f. = k − 1 = 2 and denominator d.f. = N − k = 29. For α = .01, Appendix Table IV(b) gives 5.42 as the critical value. The null hypothesis will be rejected if F > 5.42.

Computations: $\bar{\bar{x}} = \dfrac{[12(8.078) + 12(7.863) + 8(6.468)]}{32} = 7.595$

$MSTr = [12(8.078 - 7.595)^2 + 12(7.863 - 7.595)^2 + 8(6.468 - 7.595)^2]/2 = 13.822/2 = 6.911$

$$MSE = \frac{[11(1.786)^2 + 11(1.756)^2 + 7(1.339)^2]}{29} = 2.812$$

$$F = \frac{MSTr}{MSE} = \frac{6.911}{2.812} = 2.46$$

Since 2.46 < 5.42, the null hypothesis is not rejected. The data suggests that there is no difference in mean weight gain for the three salinity levels.

13.17 Let μ_1, μ_2, and μ_3 denote the true mean degree of soiling for the three mixtures.

H_0: $\mu_1 = \mu_2 = \mu_3$

H_a: At least two of the three μ_i's are different.

Test statistic: $F = \frac{MSTr}{MSE}$

Rejection region: Numerator d.f. = k − 1 = 2 and denominator d.f. = N − k = 12. For α = .05, Appendix Table IV(a) gives 3.89 as the critical value. The null hypothesis will be rejected if F > 3.89.

Computations: $\bar{\bar{x}}$ = .8873

Mixture	n	Mean	Standard Deviation
1	5	.930	.2430
2	5	.794	.0241
3	5	.938	.2430

$MSTr = [5(.93 - .8873)^2 + 5(.794 - .8873)^2 + 5(.938 - .8873)^2]/2 = .0655/2 = .0327$

$MSE = [(.243)^2 + (.0241)^2 + (.243)^2]/3 = .1188/3 = .0396$

$$F = \frac{MSTr}{MSE} = \frac{.0327}{.0396} = 0.83$$

Since 0.83 < 3.89, the null hypothesis is not rejected. The data suggests that there is not any difference in the true mean degree of soiling for the three MAA mixtures..

Section 2

13.19

Source of Variation	Degrees of Freedom	Sum of Squares	Mean Square	F
Treatments	3	75081.72	25027.24	1.70
Error	16	235419.04	14713.69	
Total	19	310500.76		

Let μ_i denote the mean number of miles to failure for brand i sparkplug (i = 1, 2, 3, 4).

H_0: $\mu_1 = \mu_2 = \mu_3 = \mu_4$

H_a: At least two of the four μ_i's are different.

Test statistic: $F = \dfrac{MSTr}{MSE}$

Rejection region: Numerator d.f. = k − 1 = 3 and denominator d.f. = N − k = 16. For α = .05, Appendix Table IV(a) gives 3.24 as the critical value. The null hypothesis will be rejected if F > 3.24.

From the ANOVA table, F = 1.70.

Since 1.7 < 3.24, the null hypothesis is not rejected. The data suggests that there is no difference between the mean number of miles to failure for the five brands of sparkplug.

13.21 a) Computations: $\bar{\bar{x}}$ = [96(2.15) + 34(2.21) + 86(1.47) + 206(1.69)]/422 = 756.1/422 = 1.792

$MSTr = [96(2.15-1.792)^2 + 34(2.21-1.792)^2 + 86(1.69-1.792)^2 + 206(1.69-1.792)^2]/3$
$= 29.304/3 = 9.768$

$MSE = \dfrac{MSTr}{F} = \dfrac{9.768}{2.56} = 3.816$

Source of Variation	Degrees of Freedom	Sum of Squares	Mean Square	F
Treatments	3	29.304	9.768	2.56
Error	418	1595.088	3.816	
Total	421	1624.392		

Let μ_i denote the mean number of hours per month absent for employees of group i (i = 1, 2, 3, 4).

H_0: $\mu_1 = \mu_2 = \mu_3 = \mu_4$

H_a: At least two of the four μ_i's are different.

Test statistic: $F = \dfrac{MSTr}{MSE}$

Rejection region: Numerator d.f. = k − 1 = 3 and denominator d.f. = N − k = 421. For α = .01, Appendix Table IV(b) gives 3.78 as the critical value. The null hypothesis will be rejected if F > 3.78.

From the ANOVA table, F = 2.56.

Since 2.56 < 3.78, the null hypothesis is not rejected. The data suggests that there is no difference between the mean number of hours per month absent for employees in the four groups.

b) From Appendix Table IV(a), the critical F value for level .05 of significance is 2.60. Since the computed F of 2.56 is less than 2.60, the P-value of this test is greater than .05.

13.23 $SSTr = 30(12.65 - 11.33)^2 + 30(10.41 - 11.33)^3 + 30(10.93 - 11.33)^2 = 82.464$

$SSE = 4468.92 + 4138.34 + 4629.44 = 13236.7$

Source of Variation	Degrees of Freedom	Sum of Squares	Mean Square	F
Treatments	2	82.464	41.232	0.27
Error	87	13236.700	152.146	
Total	89	13319.164		

Let μ_i denote true mean number of years teaching for discipline i (i = 1, 2, 3).

H_o: $\mu_1 = \mu_2 = \mu_3$

H_a: At least two of the three μ_i's are different.

Test statistic: $F = \dfrac{MSTr}{MSE}$

Rejection region: Numerator d.f. = k − 1 = 2 and denominator d.f. = N − k = 87. For α = .05, Appendix Table IV(a) gives 3.07 (approximately) as the critical value. The null hypothesis will be rejected if F > 3.07.

Since F = .27, it is less than the level .05 critical value. Hence, the null hypothesis is not rejected. The data suggests that the true mean number of years teaching is not different for the three disciplines.

13.25 a) Let μ_i denote the true mean mineral content for storage period i (i = 1, 2, 3, 4).

H_o: $\mu_1 = \mu_2 = \mu_3 = \mu_4$

H_a: At least two of the four μ_i's are different.

Test statistic: $F = \dfrac{MSTr}{MSE}$

Rejection region: Numerator d.f. = k − 1 = 3 and denominator d.f. = N − k = 20. For α = .05, Appendix Table IV(a) gives 3.10 as the critical value. The null hypothesis will be rejected if F > 3.10.

From the ANOVA table in the SAS output, F = 6.51.

Since 6.51 > 3.10, the null hypothesis is rejected. The data suggests that there is a difference in mean mineral content for at least two of the four storage periods.

b) From the ANOVA table in the SAS output, the P-value for this test is given as 0.003. The significance level α would have to be smaller than 0.003 before the null hypothesis could not be rejected.

13.27 a) $\bar{\bar{x}} = \frac{281}{18} = 15.611$, $\Sigma x = 281$, $\Sigma x^2 = 4601$

$SSTr = 6(13.83 - 15.61)^2 + 6(15 - 15.61)^2 + 6(18 - 15.61)^2 = 55.444$

$SSTo = 4601 - 18(15.611)^2 = 214.284$

$SSE = SSTo - SSTr = 214.284 - 55.444 = 158.84$

b)

Source of Variation	Degrees of Freedom	Sum of Squares	Mean Square	F
Treatments	2	55.444	27.722	2.62
Error	15	158.840	10.589	
Total	17	214.284		

c) Let μ_i denote the mean exam score for study method i (i = 1, 2, 3).

H_o: $\mu_1 = \mu_2 = \mu_3$

H_a: At least two of thc thrcc μ_i's are different.

Test statistic: $F = \frac{MSTr}{MSE}$

Rejection region: Numerator d.f. = k − 1 = 2 and denominator d.f. = N − k = 15. For $\alpha = .05$, Appendix Table IV(a) gives 3.68 as thc critical value. The null hypothesis will be rejected if F > 3.68.

The value F is read from the ANOVA table as 2.62. Since 2.62 < 3.68, the null hypothesis is not rejected. The data suggests that there is no difference between the mean exam scores. There appears to be no effect due to the technique used.

d) P-value > .05

e) No. Because the P-value > .01, the null hypothesis would not be rejected.

13.29 Since the intervals for $\mu_1 - \mu_2$ and $\mu_1 - \mu_3$ do not contain zero, μ_1 and μ_2 are judged to be different and μ_1 and μ_3 are judged to be different. Since the interval for $\mu_2 - \mu_3$ contains zero, μ_2 and μ_3 are judged not to be different. Hence, statement (iii) best describes the relationship between μ_1, μ_2, and μ_3.

13.31 a)

Source of Variation	Degrees of Freedom	Sum of Squares	Mean Square	F
Treatments	2	1835.2	917.60	35.62
Blocks	4	93.1	23.28	
Error	8	206.1	25.76	
Total	14	2134.4		

b) H_0: The mean nitrogen concentration does not depend on the rate of application.

H_a: The mean nitrogen concentration does depend on the rate of application.

Test statistic: $F = \dfrac{MSTr}{MSE}$

Rejection region: Since $k - 1 = 2$ and $(k - 1)(l - 1) = 8$, Appendix Table IV(a) gives the $\alpha = .05$ level critical value as 4.46. The null hypothesis will be rejected if $F > 4.46$.

From the ANOVA table, F = 35.62. Since 35.62 > 4.46, the null hypothesis is rejected. The mean nitrogen concentration does depend on the rate of application.

13.33 $\bar{x}_1 = 1.115,\ \bar{x}_2 = 1.5525,\ \bar{x}_3 = 1.7363,\ n_1 = n_2 = n_3 = 8$

From Exercise 13.28, MSE = .1776 with 21 degrees of freedom. $k(k - 1)/2 = 3(2)/2 = 3$.

The Bonferroni critical value is 2.60.

$$2.60\sqrt{.1776\left(\frac{1}{8} + \frac{1}{8}\right)} = .5479$$

$\mu_1 - \mu_2$: (1.1150 − 1.5525) ± .5479 = (−.9854, .1104)

$\mu_1 - \mu_3$: (1.1150 − 1.7363) ± .5479 = (−1.1692, −.0734)

$\mu_2 - \mu_3$: (1.5525 − 1.7363) ± .5479 = (−.7317, .3641)

The mean antigen concentration for asymptomatic infants differs from that of infants with late-onset meningitis. No other differences are significant.

Supplementary Exercises

13.35 $k(k-1)/2 = 6(5)/2 = 15$. The Bonferroni t critical value is 3.33.

To compare sample means based on sizes n = 4 and n = 5, use

$$3.33\sqrt{\frac{.273}{4} + \frac{.273}{5}} = 1.167$$

To compare sample means based on sizes n = 4 and n = 4, use

$$3.33\sqrt{\frac{.273}{4} + \frac{.273}{4}} = 1.23^{\cdot}$$

To compare sample means based on sizes n = 5 and n = 5, use

$$3.33\sqrt{\frac{.273}{5} + \frac{.273}{5}} = 1.10^{\cdot}$$

$\mu_1 - \mu_2$: (14.1 − 12.8) ± 1.167 = (.133, 2.467)

$\mu_1 - \mu_3$: (14.1 − 13.825) ± 1.23 = (−.955, 1.505)

$\mu_1 - \mu_4$: (14.1 − 13.1) ± 1.23 = (−.23, 2.23)

$\mu_1 - \mu_5$: (14.1 − 17.14) ± 1.167 = (−4.207, −1.873)

$\mu_1 - \mu_6$: (14.1 − 18.1) ± 1.23 = (−5.23, −2.77)

$\mu_2 - \mu_3$: (12.8 − 13.825) ± 1.167 = (−2.192, 0.142)

$\mu_2 - \mu_4$: (12.8 − 13.1) ± 1.167 = (−1.467, 0.867)

$\mu_2 - \mu_5$: (12.8 − 17.14) ± 1.10 = (−5.44, −3.24)

$\mu_2 - \mu_6$: (12.8 − 18.1) ± 1.167 = (−6.467, −4.133)

$\mu_3 - \mu_4$: (13.825 − 13.1) ± 1.23 = (−0.505, 1.955)

$\mu_3 - \mu_5$: (13.825 − 17.14) ± 1.167 = (−4.482, −2.148)

$\mu_3 - \mu_6$: (13.825 − 18.1) ± 1.23 = (−5.505, −3.045)

$\mu_4 - \mu_5$: (13.1 − 17.14) ± 1.167 = (−5.207, −2.873)

$\mu_4 - \mu_6$: (13.1 − 18.1) ± 1.23 = (−6.23, −3.77)

$\mu_5 - \mu_6$: (17.14 − 18.1) ± 1.167 = (−2.127, .207)

Brand	2	4	3	1	5	6
Sample Size	5	4	4	4	5	4
Mean	12.8	13.1	13.825	14.1	17.14	18.1

(Underlines: 12.8–13.825; 17.14–18.1; 13.1–14.1)

The mean PAPFUA levels for Mazola and Fleishmann's are the same and these two brands are different from the other four. The mean Parkay PAPFUA level differs from that of Imperial but not Chiffon or Blue Bonnet. No other differences are significant.

13.37 Let μ_i denote the mean fill weight of pocket i (i = 1, 2, 3, 4, 5).

H_0: $\mu_1 = \mu_2 = \mu_3 = \mu_4 = \mu_5$

H_a: At least two of the five μ_i's are different.

Test statistic: $F = \dfrac{MSTr}{MSE}$

Rejection region: Numerator d.f. = k − 1 = 4 and denominator d.f. = N − k = 20. For α = .05, Appendix Table IV(a) gives 2.87 as the critical value. The null hypothesis will be rejected if F > 2.87.

Computations: $\bar{\bar{x}} = 9.92$, $\Sigma x^2 = 2461$, $\bar{x}_1 = 10.08$, $\bar{x}_2 = 9.98$, $\bar{x}_3 = 9.88$, $\bar{x}_4 = 9.78$, $\bar{x}_5 = 9.88$

$SSTo = 2461 - 25(9.92)^2 = 0.84$

$SSTr = 5[(10.08)^2 + (9.98)^2 + (9.88)^2 + (9.78)^2 + (9.88)^2] - 25(.92)^2 = 0.26$

Source of Variation	Degrees of Freedom	Sum of Squares	Mean Square	F
Treatments	4	0.26	.065	2.24
Error	20	0.58	.029	
Total	24	0.84		

Since 2.24 < 2.87, the null hypothesis is not rejected. The data suggests that the mean fill weight is the same for each pocket.

13.39 Multiplying each observation in a single-factor ANOVA will change $\bar{x}_i$, $\bar{\bar{x}}$, and s_i by a factor of c. Hence, MSTr and MSE also will be changed, but by a factor of c^2. However, the F ratio remains unchanged because $c^2MSTr/c^2MSE = MSTr/MSE$. That is, the c^2 in the numerator and denominator cancel. It is reasonable to expect a test statistic not to depend on upon the unit of measurement.

13.41 The single-factor ANOVA should not be employed in this experiment. Each type of fabric should be considered a block and then the data analyzed using the analysis for a randomized block design.

13.43 $c_1 = 1$, $c_2 = -\frac{1}{2}$, $c_3 = -\frac{1}{2}$

$$c_1\bar{x}_1 + c_2\bar{x}_2 + c_3\bar{x}_3 = 44.571 - .5(45.857) - .5(48) = -2.3575$$

$$MSE\left[\frac{c_1^2}{n_1} + \frac{c_2^2}{n_2} + \frac{c_3^2}{n_3}\right] = 23.14\left[\frac{1}{7} + \frac{(-.5)^2}{7} + \frac{(-.5)^2}{7}\right] = 23.14(.2143) = 4.9586$$

t critical = 2.10

The desired confidence interval is

$$-2.3575 \pm 2.10\sqrt{4.9586} \Rightarrow -2.3575 \pm 2.10(2.2268) \Rightarrow -2.3575 \pm 4.6762 \Rightarrow (-7.03, 2.3187)$$

13.45 The mean water loss when exposed to 4 hours fumigation is different from all other means. The mean water loss when exposed to 2 hours fumigation is different from that for levels 16 and 0, but not 8. The mean water losses for duration 16, 0, and 8 hours are not different from one another. No other differences are significant.

13.47 a)

Source of Variation	Degrees of Freedom	Sum of Squares	Mean Square	F
Treatments	2	11.7	5.850	.37
Blocks	4	113.5	28.375	
Error	8	125.6	15.700	
Total	14	250.8		

b) H_0: The mean appraised value does not depend on which assessor is doing the appraisal.

H_a: The mean appraised value does depend on which assessor is doing the appraisal.

Test statistic: $F = \dfrac{\text{MSTr}}{\text{MSE}}$

Rejection region: Since $k - 1 = 2$ and $(k - 1)(l - 1) = 8$, Appendix Table IV(a) gives the $\alpha = .05$ level critical value as 4.46. The null hypothesis will be rejected if $F > 4.46$.

From the ANOVA table, $F = .37$. Since $.37 < 4.46$, the null hypothesis is not rejected. The mean appraised value does not depend on which assessor is doing the appraisal.

Chapter 14
Categorical Data and Goodness-of-Fit Tests

Section 1

14.1 a) Reject H_0 if $X^2 > 9.49$.

b) Reject H_0 if $X^2 > 13.28$.

c) Reject H_0 if $X^2 > 21.67$.

14.3 a) The critical value would be 16.27. Since the computed value of 19 exceeds the critical value of 16.27, H_0 would be rejected. The data suggests that the percentages of the four types of nuts differ from what the percentages are supposed to be.

b) If n = 40, then the chi-square test should not be used since one of the expected cell frequencies would be less than 5.

14.5 Let π_1, π_2, π_3, π_4, π_5 denote the true proportions of the various offense categories.

H_0: $\pi_1 = .307$ $\pi_2 = .386$ $\pi_3 = .093$ $\pi_4 = .206$ $\pi_5 = .008$

H_a: H_0 is not true.

Test statistic:

$$X^2 = \sum \frac{(\text{observed count} - \text{expected count})^2}{\text{expected count}}$$

Rejection region: For $\alpha = 0.05$ and d.f. = 4, reject H_0 if $X^2 > 9.49$.

Computations: n = 999

Offense	Violent crime	Crimes against property	Drug related	Public order offense	Other
Frequency	225	300	230	228	16
Expected	306.693	385.614	92.907	205.794	7.992

$$X^2 = \frac{(225 - 306.693)^2}{306.693} + \frac{(300 - 385.614)^2}{385.614} + \frac{(230 - 92.907)^2}{92.907} + \frac{(228 - 205.794)^2}{205.794} + \frac{(16 - 7.992)^2}{7.792}$$

$$= 21.760 + 19.008 + 202.294 + 2.396 + 8.024 = 253.482$$

Since 253.482 > 9.49, the null hypothesis is rejected. The data does provide sufficient evidence to conclude that the true 1989 proportions falling into the various offense categories are not the same as in 1983.

14.7 a) Let $\pi_1, \pi_2, \pi_3, \pi_4$ denote the true proportions of homicides occurring during Winter, Spring, Summer, and Fall, respectively.

H_0: $\pi_1 = \pi_2 = \pi_3 = \pi_4 = .25$

H_a: H_0 is not true.

Test statistic:

$$X^2 = \sum \frac{(\text{observed count} - \text{expected count})^2}{\text{expected count}}$$

Rejection region: For $\alpha = 0.05$, reject H_0 if the P-value is less than 0.05.

Computations: n = 1361

Season	Winter	Spring	Summer	Fall
Frequency	328	334	372	327
Expected	340.25	340.25	340.25	340.25

$$X^2 = \frac{(328 - 340.25)^2}{340.25} + \frac{(334 - 340.25)^2}{340.25} + \frac{(372 - 340.25)^2}{340.25}$$

$$+ \frac{(327 - 340.25)^2}{340.25} = .4410 + .1148 + 2.9627 + .5160 = 4.0345$$

Since the chi-square critical value for 3 degrees of freedom with level .10 is 6.25, and the calculated chi-square value is 4.03, which is smaller than the critical value, one can conclude that the P-value is in excess of .10. Since the P-value is not less than 0.05, the null hypothesis is not rejected. The data collected does not supports the conclusion that there is a difference in the proportion of homicides occurring in the four seasons.

14.9 Let π_i denote the proportion of homing pigeons who prefer direction i (i = 1, 2, 3, 4, 5, 6, 7, 8).

$$H_0: \pi_1 = \frac{1}{8}, \pi_2 = \frac{1}{8}, \pi_3 = \frac{1}{8}, \pi_4 = \frac{1}{8}, \pi_5 = \frac{1}{8}, \pi_6 = \frac{1}{8}, \pi_7 = \frac{1}{8}, \pi_8 = \frac{1}{8}.$$

H_a: H_0 is not true.

Test statistic:

$$X^2 = \sum \frac{(\text{observed count} - \text{expected count})^2}{\text{expected count}}$$

Rejection region: For d.f. = 7, Appendix Table XII gives the .10 level critical value as 12.02. The null hypothesis will be rejected if $X^2 > 12.02$.

Computations:

Direction	1	2	3	4	5	6	7	8	Total
Frequency	12	16	17	15	13	20	17	10	120
Expected	15	15	15	15	15	15	15	15	

$$X^2 = \frac{(12-15)^2}{15} + \frac{(16-15)^2}{15} + \frac{(17-15)^2}{15} + \frac{(15-15)^2}{15}$$

$$+ \frac{(13-15)^2}{15} + \frac{(20-15)^2}{15} + \frac{(17-15)^2}{15} + \frac{(10-15)^2}{15} = \frac{72}{15} = 4.8$$

Since 4.8 < 12.02, the null hypothesis is not rejected. The data supports the hypothesis that when homing pigeons are disoriented in a certain manner, they exhibit no preference for any direction of flight after take-off.

Section 2

14.11 a) d.f. = (4 – 1)(5 – 1) = 12. Reject H_0 if $X^2 > 21.03$.

b) At $\alpha = .10$, H_0 would be rejected if $X^2 > 18.55$. With $X^2 = 7.2$, then the null hypothesis would not be rejected. The data does not support the conclusion that educational level and preferred candidate are dependent factors.

c) d.f. = (4 – 1)(4 – 1) = 9. Reject H_0 if $X^2 > 16.92$. With $X^2 = 14.5$, then the null hypothesis would not be rejected. The data does not support the conclusion that educational level and preferred candidate are dependent factors.

14.13 H_0: Job satisfaction and teaching level are independent.

H_a: Job satisfaction and teaching level are dependent.

Test statistic:

$$X^2 = \sum \frac{(\text{observed count} - \text{expected count})^2}{\text{expected count}}$$

Rejection region: For d.f. = 2, Appendix Table XII gives the .05 level critical value as 5.99. The null hypothesis will be rejected if $X^2 > 5.99$.

Computations:

		Job satisfaction		
		Satisfied	Unsatisfied	
	College	74 (63.763)	43 (53.237)	117
Teaching level	High School	224 (215.270)	171 (179.730)	395
	Elementary	126 (144.967)	140 (121.033)	266
	Total	424	354	778

$$X^2 = \frac{(74 - 63.763)^2}{63.763} + \frac{(43 - 53.237)^2}{53.237} + \frac{(224 - 215.270)^2}{215.270}$$

$$+ \frac{(171 - 179.730)^2}{179.730} + \frac{(126 - 144.967)^2}{144.967} + \frac{(140 - 121.023)^2}{121.023}$$

$$= 1.644 + 1.968 + .354 + .424 + 2.482 + 2.972 = 9.844$$

Since the calculated chi-square value of 9.844 exceeds the chi-square critical value of 5.99, H_0 is rejected. The data supports the conclusion that there is an association between job satisfaction and teaching level.

14.15 a) H_0: Proportions of parasitized and nonparasitized fish are the same for all three treatments.

H_a: H_0 is not true.

Test statistic:

$$X^2 = \sum \frac{(\text{observed count} - \text{expected count})^2}{\text{expected count}}$$

Rejection region: For d.f. = 2, Appendix Table XII gives the .01 level critical value as 9.21. The null hypothesis will be rejected if $X^2 > 9.21$.

Treatment	Parasitized	Nonparasitized	Total
Control	30 (23.0)	3 (10.0)	33
Old oil	16 (16.7)	8 (7.3)	24
New oil	16 (22.3)	16 (9.7)	32
Total	62	27	89

$$X^2 = \frac{(30 - 23.0)^2}{23.0} + \frac{(3 - 10.0)^2}{10.0} + \frac{(16 - 16.7)^2}{16.7}$$

$$+ \frac{(8 - 7.3)^2}{7.3} + \frac{(16 - 22.3)^2}{22.3} + \frac{(16 - 9.7)^2}{9.7}$$

$$= 2.130 + 4.900 + 0.029 + 0.067 + 1.780 + 4.092 = 12.998$$

Since the calculated chi-square value of 12.998 exceeds the chi-square critical value of 9.21, H_0 is rejected. The data indicates that the proportions of parasitized and nonparasitized fish are not the same for all three treatments.

b) For 2 degrees of freedom, the .01 chi-square value is 9.21 and the .001 chi-square value is 13.82. Since the calculated value is between 9.21 and 13.82, one can conclude .01 > P-value > .001.

14.17 a) This problem involves comparing two populations. The design of the experiment was such that two random samples were taken, one from users of each screen type. Therefore, the marginal row totals are known.

b) H_0: No difference between users of screen type with respect to the proportion falling into the two reflection categories.

H_a: Users of screen type differ with respect to the proportion falling into each of the reflection categories.

Test statistic:

$$X^2 = \sum \frac{(\text{observed count} - \text{expected count})^2}{\text{expected count}}$$

Rejection region: For d.f. = (2 – 1)(2 – 1) = 1, Appendix Table XII gives the .05 level critical value as 3.84. The null hypothesis will be rejected if $X^2 > 3.84$.

	No	Yes	Total
Nonadj.	15 (21.34)	50 (43.66)	65
Adj.	28 (21.66)	38 (44.34)	66
Total	43	88	131

$$X^2 = \frac{(15 - 21.34)^2}{21.34} + \frac{(50 - 43.66)^2}{43.66} + \frac{(28 - 21.66)^2}{21.66} + \frac{(38 - 44.34)^2}{44.34}$$

$$= 1.88 + .92 + 1.86 + .91 = 5.57$$

Since 5.57 > 3.84, the null hypothesis is rejected. The data suggests that users of the two screen types differ with respect to the proportions falling into the no and yes categories of annoying reflections.

c) The following test statistic could be used.

$$z = \frac{(p_1 - p_2)}{\sqrt{\frac{p_c(1 - p_c)}{n_1} + \frac{p_c(1 - p_c)}{n_2}}}$$

The X^2 statistic can be used only to detect differences. The z statistic can be used to test for differences, as well as the direction of difference. That is, the z statistic can be used on both one-sided and two-sided tests, but the X^2 statistic can be used only for two-sided tests. Therefore, if the researchers were interested in determining whether the proportion experiencing annoying reflections was smaller for the adjustable VDT screens, they would use the z statistic.

14.19 H_0: Response and gender are independent variables.

H_a: Response and gender are dependent variables.

Test statistic:

$$X^2 = \sum \frac{(\text{observed count} - \text{expected count})^2}{\text{expected count}}$$

Rejection region: For d.f. = (2 – 1)(4 – 1) = 3, reject H_0 at level of significance .01, if $X^2 > 11.34$.

Computations:

Emotion	Gender: Male	Gender: Female	Total
Anger	34 (31.925)	27 (29.075)	61
Pain	28 (23.551)	17 (21.449)	45
Happiness	12 (16.224)	19 (14.776)	31
Love	38 (40.299)	39 (36.701)	77
Total	112	102	214

$$X^2 = \frac{(34 - 31.925)^2}{31.925} + \frac{(27 - 29.075)^2}{29.075} + \frac{(28 - 23.551)^2}{23.551}$$

$$+ \frac{(17 - 21.449)^2}{21.449} + \frac{(12 - 16.224)^2}{16.224} + \frac{(19 - 14.776)^2}{14.776}$$

$$+ \frac{(38 - 40.299)^2}{40.299} + \frac{(39 - 36.701)^2}{36.701}$$

$$= .1349 + .1481 + .8406 + .9228 + 1.0997 + 1.2075 + .1312 + .1440 = 4.629$$

Since $4.629 < 11.34$, the null hypothesis is not rejected. The data does not support the conclusion that there is an association between the two variables response and gender.

Supplementary Exercises

14.21 H_o: proportions in each response category are the same for graphic and plain cover designs.

H_a: H_o is not true.

Test statistic:

$$X^2 = \sum \frac{(\text{observed count} - \text{expected count})^2}{\text{expected count}}$$

For level of significance .05, reject Ho if $X^2 > 9.49$.

The expected cell count for graphic cover, 1-7 day response is:

$$\text{expected count} = \frac{154(414)}{841} = 75.8$$

The expected cell count for plain cover, 15-31 day response is:

$$\text{expected count} = \frac{102(427)}{841} = 51.8$$

The other expected cell counts are found in an analogous fashion. The expected cell counts appear in parentheses in the following table.

Computations:

Response		1 - 7	8 - 14	15 - 31	32 - 60	Not Returned	Total
Cover	Graphic	70 (75.8)	76 (63.5)	51 (50.2)	19 (25.1)	198 (199.4)	414
	Plain	84 (78.2)	53 (65.5)	51 (51.8)	32 (25.9)	207 (205.6)	427
	Total	154	129	102	51	405	841

$$X^2 = \frac{(70 - 75.8)^2}{75.8} + \frac{(76 - 63.5)^2}{63.5} + \frac{(51 - 50.2)^2}{50.2} + \frac{(19 - 25.1)^2}{25.1} + \frac{(198 - 199.4)^2}{199.4}$$

$$X^2 = \frac{(84 - 78.2)^2}{78.2} + \frac{(53 - 65.5)^2}{65.5} + \frac{(51 - 51.8)^2}{51.8} + \frac{(32 - 25.9)^2}{25.9} + \frac{(207 - 205.6)^2}{205.6}$$

$$= .4438 + 2.4606 + .0127 + 1.4825 + .0098 + .4302 + 2.3855 + .0124 + 1.4367 + .0095 = 8.68$$

Since 8.68 does not exceed 9.49, the null hypothesis is not rejected. The data does not support the conclusion that the proportions falling in the various response categories differ for the two cover designs.

14.23 H_0: The true proportions of red, yellow and white are 9/16, 3/16, and 4/16, respectively.

H_a: At least one proportion not as specified in H_0.

Test statistic:

$$X^2 = \sum \frac{(\text{observed count} - \text{expected count})^2}{\text{expected count}}$$

Rejection region: For d.f. = k − 1 = 2, Appendix Table XII gives the .10 level critical value as 4.61. The null hypothesis will be rejected if $X^2 > 4.61$.

Computations:

	Red	Yellow	White	Total
Observed	195	73	100	368
Expected	207	69	92	368

$$X^2 = \frac{(195 - 207)^2}{207} + \frac{(73 - 69)^2}{69} + \frac{(100 - 92)^2}{92} = .70 + .23 + .70 = 1.63$$

Since 1.63 < 4.61, the null hypothesis is not rejected. The data does not cast doubt on the appropriateness of this genetic theory.

14.25 H_0: Age and rate believed attainable are independent variables.

H_a: Age and rate believed attainable are dependent variables.

Test statistic:

$$X^2 = \sum \frac{(\text{observed count} - \text{expected count})^2}{\text{expected count}}$$

Rejection region: For d.f. = (4 − 1)(4 − 1) = 9, Appendix Table XII gives the .01 level critical value as 21.67. The null hypothesis will be rejected if $X^2 > 21.67$.

Computations:

		Rates believed attainable				
		0 - 5	6 - 10	11 - 15	Over 15	Total
	Under 45	15 (28.4)	51 (72.1)	51 (28.2)	29 (17.3)	146
	45 - 54	31 (54.8)	133 (139.3)	70 (54.5)	48 (33.4)	282
Age	55 - 64	59 (49.2)	139 (124.9)	35 (48.9)	20 (29.9)	253
	65 over	84 (56.6)	157 (143.7)	32 (56.3)	18 (34.4)	291
	Total	189	480	188	115	972

$$X^2 = \frac{(15 - 28.4)^2}{28.4} + \frac{(51 - 72.1)^2}{72.1} + \dots + \frac{(18 - 34.4)^2}{34.4}$$

$$= 6.31 + 6.17 + 18.35 + 7.96 + 10.36 + .28 + 4.38 + 6.42 + 1.95$$

$$+ 1.58 + 3.97 + 3.3 + 13.28 + 1.23 + 10.48 + 7.84$$

$$= 103.87$$

Since 103.87 > 21.67, the null hypothesis is rejected. The data very strongly suggests that the variables, age and rate believed attainable are dependent variables, so there is association.

14.27 H_0: The true proportions of individuals in each of the cocaine use categories do not differ for the three treatments.

H_a: The true proportions of individuals in each of the cocaine use categories differ for the three treatments.

Test statistic:

$$X^2 = \sum \frac{(\text{observed count} - \text{expected count})^2}{\text{expected count}}$$

Rejection region: For d.f. = (4 – 1)(3 – 1) = 6 and Appendix Table XII gives the .05 level critical value as 12.59. The null hypothesis will be rejected if $X^2 > 12.59$.

Computations:

	Treatment			
Usage	A	B	C	Total
None	149 (118.9)	75 (84.8)	8 (28.3)	232
1 - 2	26 (34.8)	27 (24.9)	15 (8.3)	68
3 - 6	6 (19.0)	20 (13.5)	11 (4.5)	37
7 or more	4 (12.3)	10 (8.8)	10 (2.9)	24
Total	185	132	44	361

$$X^2 = \frac{(149 - 118.9)^2}{118.9} + \frac{(75 - 84.8)^2}{84.8} + \dots + \frac{(10 - 2.9)^2}{2.9}$$

$$= 7.62 + 1.14 + 14.54 + 2.25 + .18 + 5.44 + 8.86 + 3.1 + 9.34 + 5.6 + .17 + 17.11 = 75.35$$

Note: There are two cells with expected counts of 5 or less. If you combine the two usage categories of 3 - 6 and 7 or more the following analysis results.

	Treatment			
Usage	A	B	C	Total
none	149 (118.9)	75 (84.8)	8 (28.3)	232
1 - 2	26 (34.8)	27 (24.9)	15 (8.3)	68
3 or more	10 (31.3)	30 (22.3)	21 (7.4)	61
Total	185	132	44	361

$$X^2 = \frac{(149 - 118.9)^2}{118.9} + \frac{(75 - 84.8)^2}{84.8} + \ldots + \frac{(21 - 7.4)^2}{7.4}$$

$$= 7.62 + 1.14 + 14.54 + 2.25 + .18 + 5.44 + 14.46 + 2.65 + 24.75 = 73.03$$

The d.f. = (3 − 1)(3 − 1) = 4 and the .05 level critical value is 9.49.

In either analysis, the null hypothesis is rejected. The data suggests the true proportions of individuals in each of the cocaine use categories differ for the three treatments.